CW00567216

# How to Write
# Biological Sciences

# How to Write a PhD in Biological Sciences

## Biological Sciences
### A Guide for the Uninitiated

John Measey

## CRC Press
Taylor & Francis Group

Boca Raton London New York

CRC Press is an imprint of the
Taylor & Francis Group, an **informa** business

First edition published 2022
by CRC Press
6000 Broken Sound Parkway NW, Suite 300, Boca Raton, FL 33487-2742

and by CRC Press
2 Park Square, Milton Park, Abingdon, Oxon, OX14 4RN

© 2022 John Measey

CRC Press is an imprint of Taylor & Francis Group, LLC

Reasonable efforts have been made to publish reliable data and information, but the author and publisher cannot assume responsibility for the validity of all materials or the consequences of their use. The authors and publishers have attempted to trace the copyright holders of all material reproduced in this publication and apologize to copyright holders if permission to publish in this form has not been obtained. If any copyright material has not been acknowledged please write and let us know so we may rectify in any future reprint.

Except as permitted under U.S. Copyright Law, no part of this book may be reprinted, reproduced, transmitted, or utilized in any form by any electronic, mechanical, or other means, now known or hereafter invented, including photocopying, microfilming, and recording, or in any information storage or retrieval system, without written permission from the publishers.

For permission to photocopy or use material electronically from this work, access www.copyright. com or contact the Copyright Clearance Center, Inc. (CCC), 222 Rosewood Drive, Danvers, MA 01923, 978-750-8400. For works that are not available on CCC please contact mpkbookspermissions@tandf.co.uk

*Trademark notice:* Product or corporate names may be trademarks or registered trademarks and are used only for identification and explanation without intent to infringe.

ISBN: 978-1-032-08021-5 (hbk)
ISBN: 978-1-032-08020-8 (pbk)
ISBN: 978-1-003-21256-0 (ebk)

DOI: 10.1201/9781003212560

Typeset in CMR10 font
by KnowledgeWorks Global Ltd.

**Cover illustration: Saint George and the Dragon was adapted by the author from an original work by Paolo Uccello (~1470). While the student (St George), supported by their laboratory (White Horse), is attempting to bring about the end of their PhD (Dragon) using this Fool's Guide (lance), the advisor (Princess) is leading the PhD by a leash from the darkness into the community towards the Scientific Project (town).**

**The follow-up to this book, *How to Publish in Biological Sciences: A Guide for the Uninitiated*, is available at www.routledge.com/9781032116419**

To my students,

past, present and future

# Contents

# List of Tables

# List of Figures

# *Welcome*

Welcome to this guide on how to write a PhD in Biological Sciences. I will not tell you what to write. My intention is to tell you how to approach writing with respect to style, format and formula (planning).

Writing a PhD is a daunting prospect. One does not put pen to paper, or, rather, finger to keyboard, only after finishing all of your experiments and/or labwork. Writing is actually something you should start from day one of your PhD, or even before day one, when you make the first initial pitch of big ideas to your advisor on what you want to do. If you can find the discipline to start writing from day one, then you are much more likely to make a better go of it than waiting until the final, pressured stretch. But to do this, you need to understand how to write a PhD thesis. You need to understand what you are aiming to achieve, and, crucially, you need to understand why you are doing your PhD.

This guide intends to demystify the writing process so that the eventual goal is clear, and to provide the necessary information on how to get there. It takes you from those initial ideas, all the way through proposal writing, posing hypotheses, testing them and then how to write it all up.

This book started as a series of blog posts written by me for my students (john.measey.com/blog). I wrote them because I felt that I was repeating the same advice year after year. Instead of rewriting comments on manuscripts time after time, I decided to consolidate them and put them all in one place. Later, I started looking for particular problems that students were having, and then writing blog posts to try to preempt those problems and to provide solutions before tasks were completed. Finally, I have tried to fill in the missing gaps in the blog posts. But there are still plenty of gaps, and I regard this book as a work in progress. It is my intention that this book will always be added to, and improved by, a community of practice, and I hope that this will include you.

## Why read this book

Embarking on a PhD is intimidating as, for most students, it will be their first experience working within the academic system. The voyage of discovery is often made very frustrating as much of what goes on in academia is assumed knowledge. Academics accumulate knowledge throughout their careers, but what can be done for those who are uninitiated? What is needed is a guide that postgraduate students can refer to before, during and while making decisions about their time within academia. Note that this is not a rulebook. There are times when the guide will be accurate and others when it will be vague but still provide some insight to point you in directions where you can explore more. The intention then is to provide you with a starting point from which you can establish your confidence in the academic writing process and build your own creativity.

## Structure of the book

This book is intended to be a guide. Don't sit down and read it from cover to cover. If you do, then I think you have likely misunderstood what this guide is for. Instead, I want you to think about this book as a guide much in the same way that you might pick up a guide to a foreign city, country or even like Douglas Adams' fabled electronic guide *The Hitchhiker's Guide to the Galaxy* (2017) (if you are not familiar, see here[1]). However, instead of having hostels or restaurants recommended by name with a description of what makes them particularly praiseworthy, this guide provides a generic idea of what such places are for, what they tend to be like, the best ways to get to them and pitfalls to avoid.

None of the topics is explored in any great depth. This is deliberate as this guide is here to provide you with a starting point for you to explore if you think it is necessary. I have provided literature (where possible), and you should use this in the standard way to research the topic yourself should you need or want to know any more. In the same way, none of the advice provided in this book should override what you decide for yourself or with your advisor. Make your decisions from an informed position.

---

[1] https://en.wikipedia.org/wiki/The_Hitchhiker%27s_Guide_to_the_Galaxy

**The book is divided into four main parts:**

**Part I** – Part one is written as something that you should read early on in the process, once you've committed to doing your PhD. This section of the book deals with what a PhD is, how it's examined and with whom, where and when to do it. Perversely, it asks you what you want to do after you've finished, pointing out that you will have opportunities to open doors during the process. At the start of your PhD is the time to be thinking of the bigger questions both in your study area and your life, and this section provides some pointers about things to consider. Working within academia is a particularly stressful environment, and students should be aware of the state of their mental health, how to be aware if it is deteriorating and what you can do to help yourself. I also encourage you to participate in the **scientific project** and to become aware of your role as a scientist within society.

**Part II** – Part two of the book sets out a practical guide to getting started with writing. The approach I've taken here is describing the production of a PhD in Biological Sciences which is made up of data chapters, and where each chapter is essentially a potential publication in a science journal. If you didn't want to do a PhD but just wanted to write a publication, then most of what is here (and in Part III) will be of use as you could take the same formulaic approach as given in this book. Similarly, if you were undertaking an MSc by research, this section would also be useful. The approach taken is 'hypothesis-centric', and how the entire chapter will be built around answering the hypothesis posed. This section explains how the hypothesis and all logical arguments that you make in the text are constructed from existing literature – **standing on the shoulders of giants** – with a practical guide on how to use citations. In addition to some nuts and bolts of how to write paragraphs, construct arguments, remain concise and avoid plagiarising, this section attempts to cover a lot of technical aspects of writing for the biological sciences.

**Part III** – A typical PhD will consist of five data chapters where each data chapter is a publication. The standard formula is to write this in sections consisting of an introduction, methods and materials, results and a discussion. Together with the title, abstract, references and supplementary information, all of these parts also make up most scientific publications. In this section of the book, I go through each of these components using a formulaic approach. To get you started writing, I suggest how you can build each section from an outline that is fleshed out with literature and works around a hypothesis (from Part II). In addition to the sections needed for your thesis, I also cover parts needed to submit a paper, including the title page, abstract, supplementary material, authorship and acknowledgements. Therefore, this part of the book will be relevant for anyone setting out to write a scientific paper in the biological sciences, especially if you are having trouble getting started.

**Part IV** – The last and shortest section of this book talks about what you still need to do once your data chapters are written. For most PhD theses, you will need to write a short introductory chapter (sometimes a literature review) that comes before these data chapters, and a concluding synthesis chapter at the end. These chapters make your thesis into a 'body of work', and allow you space to reflect on what you have achieved in your PhD studies. I have framed their content at the bare minimum required, as many students will find themselves under a great deal of pressure to submit their thesis to a prescribed deadline. I also provide some alternative approaches for those who have the luxury of more time to attack these sections in more detail. The very end of the book sets out some reasons for why you need to publish your thesis chapters.

Throughout this guide you will find links and references to things to read elsewhere. The reality is that there is a lot out there for you to read. I am not the only person to write a guide, and I can't pretend that all the answers are here in this book. You will find that in your particular situation you need to look around. What I aim to do in this book is provide you with enough background information to help in your onward search, and provide you with enough inside knowledge to do a PhD.

## Why 'A Guide for the Uninitiated'?

I think that most people with doctorates would agree that a PhD is not awarded to people because they are particularly bright or smart. If you had to be a genius, then I wouldn't have a PhD. Indeed, I don't consider myself to be particularly clever, but I worked very hard to get my PhD. I was hampered by the fact that I didn't know anything about the goals and aims of the academic process of working towards a PhD, so it took a lot more work, wasted time and (let's not mince our words) real pain. The end product was a fraction of the potential that I could have achieved if I had understood more about the process. If I had only had a guide to tell me what it was all about, I could have saved myself so much time and energy. In short, I feel that I was uninitiated, and this is the guide I wish that I had had.

So, this guide is my practical attempt to help you to get the most that you can out of your PhD time. After all, it should be an amazing journey that culminates with you achieving academic maturity: the ability to conceive and carry out scientific investigations. When you have a PhD, you become a scientist. Too often, however, it's a nightmare journey of cul-de-sacs and groping in the dark.

You should be aware from the outset that doing a PhD is a major undertaking. It will take several years of your life and will impact heavily on your day-to-day existence, to the extent that your mental health may suffer. A PhD represents the culmination of an incredible amount of work, not to mention that you've already had to complete one to three previous degrees beforehand. It may not be something that you ever finish to your own satisfaction. But you must finish it.

## Acknowledgements

Before I go on, there are a great many people that I need to thank. First and foremost are my students, past and present, who have inspired me to put together first the blog posts and then the book. It is because you wanted more that I put this together. I have also been a student, and have been inspired by colleagues around the world who are exemplary advisors. This book contains lots of links to blogs and articles written and posted freely on the internet by others who also aim to demystify and help. I thank this greater academic community (especially #academicTwitter) for sharing and inspiring. Thanks go to the many reviewers and editors who have taken their time to improve my writing. I am still learning. Lots of the text in this book has been improved by feedback from my students and postdocs. A special mention must go to my brother, Richard, who has hosted my lab website for more than a decade, and especially for saving blog posts from hacking attacks. Thanks also to my wife, Thalassa, who proofread many of the blog posts after I had published them late at night so that I could correct them over breakfast in the morning. James Baxter-Gilbert, Jack Dougherty, Anthony Herrel, Allan Ellis, Andrea Melotto, Joyce Moss, Lisa and Mark O'Connell, Claire Riss, Pat Schloss, James Vonesh and Carla Wagener all read or commented on different aspects of the book. Thanks are also due to my colleagues at the Centre for Invasion Biology, the Department of Botany and Zoology and in the library at Stellenbosch University.

John Measey
Cape Town

# *About the Author*

John Measey is Associate Professor of Biological Sciences at Stellenbosch University. He has authored or co-authored more than 200 peer-reviewed scientific papers and book chapters, and five books. He has been the Editor-in-Chief of an ISI journal for nine years, and currently serves as Associate Editor for four other journals. He has graduated more than 20 postgraduate students, and his blog on writing and publishing in biological sciences is read by thousands globally. British born and educated, he lives and works in the beautiful Western Cape, South Africa.

## Do you have something to contribute?

This book is written in bookdown (Xie, 2016) specifically to make it a 'live project' that will be open to anyone who wants to contribute, improve or use as the basis for their own book. The easiest way for readers to contribute content directly is through a GitHub pull request[2]. At the repository for this book, you will find Rmd files for each chapter, and as a GitHub user, you can simply edit the Rmd files and submit the changes. If I am happy with the changes proposed, I will merge your content with that of the book and add your name to the Acknowledgements.

One of the amazing potentials for bookdown books[3] is that all the files for this book are hosted in a repository on Github[4]. You have the opportunity to fork this repository and write your own version for a different discipline, a different language or for a different region of the world. It is also my hope that this guide can grow to become a community of practice for those conducting PhDs in Biological Sciences. It will not be possible to cover every aspect of writing a PhD in Biological Sciences, but it may be that I have missed out on ones that are very important to you. Equally, parts of what is currently written will become obsolete as new initiatives begin and old problems are resolved. For this reason, this guide needs to be a 'living document', and anyone who wants to provide feedback or contribute new sections is more than welcome.

---

[2] https://help.github.com/articles/about-pull-requests/

[3] https://bookdown.org/

[4] https://github.com/johnmeasey/How-to-write-a-PhD-in-Biological-Sciences/tree/main

# Part I

# Right from the Very Start

DOI: 10.1201/9781003212560-1

# 1

## Introduction

Right from the very start of your PhD, it is important for you to become familiar with the goal of what you are trying to achieve, how you plan to achieve it, and (believe it or not) what you are going to do afterwards. There are a lot of things to think about as you start your PhD, and for many of these it is far easier to make the informed decisions from the outset. For example, reconciling yourself to using the statistical package R will be much easier from the start, rather than only realising under pressure at the end of your PhD that there is no way that you can finish unless you take it on. Getting, and staying in, the right mental framework will also be very important for your ability to work under pressure. Again, getting into good habits at the beginning will put you in a commanding position later.

## 1.1 So you are doing a PhD?

In this first chapter of the book, I intend to make you familiar with what a PhD is, how to approach thinking about it, and what to look for in an institution and advisor. You may already have some or all of these things fixed, but there is still some use in going over these basic points. Try to keep an open mind. Doing a PhD will take several years of your life, and may well define your future career prospects. Therefore, to say that this is a big decision is an understatement. It is not a substitute for getting a job or a way of remaining a student for longer. If after reading this section you are in any doubt about whether or not a PhD is right for you, then I would suggest that you do need to think about it further. Don't take my word for it, go and talk to people who have done a PhD (whether or not they now have a job), or those who are doing a PhD now. Especially if their experience is in your focal research area, they are likely to know a lot more about it.

DOI: 10.1201/9781003212560-2

## 1.1.1   What is a PhD?

PhD comes from the Latin *Philosophiae Doctor*, and means "Doctor of Philosophy". All doctorates derive from philosophy which is essentially the mother of all academic subjects.

Obtaining a PhD allows you to (legitimately) call yourself a Doctor, and/or put the letters PhD after your name. You likely already have other letters (BSc, MSc, etc.) after your name, so this won't be anything new. Becoming a Doctor is (in most parts of the world) the last step in your formal training as a researcher. However, in some countries (see list here[1]) a further process (Habilitation) is required to lead research groups and/or teach at universities (leading the way to a professorship).

PhDs are granted by universities (for the most part) and those are the institutions that set the standards. In many countries, these universities are required to meet the criteria of national bodies. Essentially, the top-down control is to ensure that the PhD qualification adheres to national and international standards. You may be aware that there are mutterings of whether qualifications are what they used to be. This has also been said of PhDs. However, I'm of the opinion that the standards are much as they were. You will really set the level on the bar of your own PhD. You either pass or fail. If your intention were to do the 'bare minimum' required to get a PhD, then you would have to look carefully for an advisor that would be happy to preside over such a thesis. Suffice it to say that this would be a very poor place to start, and there may be lots of other reasons why doing a PhD really is not appropriate for you.

## 1.1.2   Is a PhD more than a thesis?

Yes. The thesis is really only the final evidence of all of the work and experience that you undertake while doing your PhD. The time that you spend doing your PhD will likely include participating in conferences, giving talks, participating in journal clubs, seminars, workshops and working groups. In addition to all of the writing skills, on which this book focuses, you will develop skills in techniques that relate to your particular subject area. You will also learn and develop your skills in presentation, analysis and treatment of data. It is an intense learning experience. It should also be a fun and positive experience.

While your PhD thesis will be a crucial and focal product of your PhD, and producing it will require a large amount of work, the thesis cannot be considered equivalent to the PhD.

---

[1] https://en.wikipedia.org/wiki/Habilitation

Having said all of this, we have not really answered the question:

## What is a PhD?

Because every PhD is different, it's quite a hard question to answer. Instead, I think that a better approach is to ask for the qualities that examiners are asked to assess when examining your PhD thesis. Again, this will be different in different institutions and countries. Some countries require an oral defence where a specialist jury is presented with the PhD dissertation and can grill the candidate for hours in public. Others, like my defense at Bristol University, happened behind closed doors with just two specialist examiners: an internal and an external.

### 1.1.3 Criteria used to judge PhD studies

In order to provide you with some general idea of what is required, I list below the (edited) questions asked of examiners for a PhD thesis by my university (I encourage you to ask for the requirements at your own institution):

- Has the motivation for the objectives of the study been formulated satisfactorily?
- Do the research results constitute a meaningful contribution to the knowledge of and insight into the relevant field of study?
- Does the dissertation distinguish clearly between own, new contributions to, and known results in the relevant field of study?
- Is the candidate capable of evaluating the scientific meaning of their results and of placing this in context within existing knowledge in the field of study?
- Does the candidate show signs of independent, critical thinking or other signs of originality?
- Does the candidate show that they are sufficiently capable of doing independent research?
- Does the dissertation show that the candidate is sufficiently familiar with the relevant research techniques and methods?
- Does the dissertation show conversance with, and a critical attitude towards the pertinent literature?
- Is the material presented in a clear, systematic and logical manner?
- Is the linguistic, stylistic and technical editing of the dissertation acceptable?
- Are the research results acceptable for publication?

If you compare these PhD requirements with those of an MSc, the key difference is that for a PhD you are required to produce a "meaningful contribution to the knowledge of and insight into the relevant field of study", and show that you are "...capable of doing independent research". Thus, a PhD degree prepares you to be an independent researcher, and so you are expected to have a critical attitude towards other research.

## 1.1.4   Who are the examiners?

The next question to ask is who the examiners are? Again, different institutions will have different guidelines about who can be an examiner for a PhD. It is likely that most will agree that they should be holders of PhDs, in tenured positions (sometimes universities insist on full professorial appointments for examiners), and with experience of advising PhD students as well as having a proven record of their own publications. Hence, most PhD examiners will likely be senior academic figures. Their views are likely to have been informed by their own experience in academia, their own institutions and systems and prior experience of examining PhDs at other institutions.

I can tell you that many examiners will treat the exercise the same irrespective of the questions asked of them. They will treat the examination as an extra-large task in peer review. This is not unfair. Personally, I think that there are some special extra hurdles required for a PhD (over and above five chapters that are publishable):

- First is the clear ability to conceptualise and carry out a body of scientific work that is novel. Here I mean novel in that it has not simply been a repeat of some previous work on a different taxon or system. This is easier to judge from an oral defense, over reading the thesis.
- Second is the clear knowledge of how the work that is conducted fits into the bigger picture. That is, that the work demonstrates that the person is a scientist at a broader level than the narrow focus of their research.
- Third is whether the student has been able to bring the study together in a way that shows that it is more than the sum of its parts. In essence, together the five data chapters should provide a bigger overview of a broader topic, and show the direction of future work.

I prefer a thesis that fits together within a stated framework. I like PhD students to conceptualise how their chapters are interlinked, and to present this as a figure in the introduction of their proposal. A useful concept to be aware of is the 'Hierarchy of Hypotheses' (Heger and Jeschke, 2018). In this approach, you are encouraged to consider the 'bigger question' in your research area, identifying both what studies have produced evidence and identifying gaps from a theory driven approach. Generally, it is only possible to consider these types of concepts once you are familiar with the literature (both theoretical and experimental).

## 1.1.5   What will you do after your PhD?

It sounds silly to think about what comes after your PhD even as you are just starting, but the reality is that you always need to be planning ahead in order to make the most of what you are doing now. During your PhD you will have lots of opportunities in many different areas of academia, teaching and practically applied areas. Knowing what you want to do when you finish will help guide your decisions so that you can maximise opportunities as they arise. Both inside and outside of academia, it is likely to be the contacts that you make that will help in your next career move.

### 1.1.5.1 Doing a PhD actually reduces your chance of employment and lowers your pay

The reality is that most employers don't want someone with a PhD, because it is a very academic qualification. It will take you so far down an academic rabbit hole that for an employer there's very little to use. Most employers want people who understand the nuts and bolts of doing research, but who can apply this knowledge outside of the academic system. Employers who want people with PhDs (and there are a few) generally want professional researchers who are essentially academics. Good work if you can get it, but it tends not to pay well. You might end up only being fit to be an academic, and that life is getting increasingly harder without any end to the expectations of the employers, the equivalent of doing a PhD every year. Something only very few people will enjoy.

At the same time, governments the world over, and especially developing countries, are providing bursaries for people to do PhDs. Why? The number of PhDs in a country is considered (by the World Bank[2]) to be one of many measures of the development of that country. As the percentage gets larger, so countries can convince the World Bank that they are developing (and then get granted its favours, see Figure 1.1). This is all well and good, but strategically, governments should consider why they want these researchers in their populations, and more particularly for what kinds of jobs. It seems somewhat unfair to produce lots of young PhD holders that have no careers to join, or prospects to use their PhD just so that your government can tick a check box on their road to development, and so borrow more money from the World Bank. If you are from a developing country, you cannot assume that just because you are given a bursary by your government it follows that they will give you a job or a career. This will likely always be your own task.

In a recent report from the UK, it was found that by age 35, the earnings gap for those with PhDs was lower than the same cohort that left academia after their BSc or MSc (Britton et al., 2020). Women who obtain PhDs tend to increase their earnings by ~8%, while men *reduce* their earnings by ~9%. However, for those in biological sciences women seem likely to earn no more than those with undergraduate degrees by age 35. Men earn around 10% less. The point here is that you should not be doing a PhD in Biological Sciences because you think you will earn more money in your following career. Having

---

[2]https://datacatalog.worldbank.org/dataset/world-development-indicators

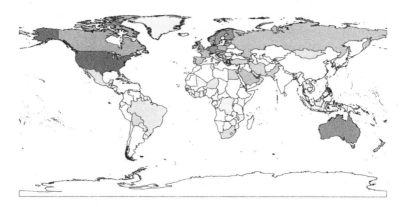

**FIGURE 1.1: Where in the world are there most PhDs?** In this map of the world, countries shaded are those that have data declaring the proportion of PhD (or equivalent) in their population (max percentage between 2010 and 2020 data from the World Bank). This ranges from 2.97% in Switzerland (darkest) to 0.02% in DR Congo (lightest shading). Notably, there is no data held in these years for many countries in Asia, Africa and South America.

made this point, I also feel that it is important to point out that how much you earn may not be directly related to job satisfaction or lifetime happiness. To my knowledge, this is not (yet) reported, but could be a lot higher for those with PhDs, even though they earn less.

It is important to realise that many academics are looking for good students to do PhD work. This is because they need someone to do research within their groups, and students or postdocs do most of this work. For this reason, an academic you know while you're doing your MSc studies may start to try to persuade you to do a PhD. This could be fantastic for you, it could be just what you're looking for. But the reality is that you will never do a PhD for somebody else, you are always going to be doing it for yourself. If the only reason you are doing a PhD is because someone else told you that you should then you are likely to have a really hard time. By the time you graduate, you may have a PhD but without any idea of what you're going to do with it.

### 1.1.6 Think about your career before you start your PhD

The last thing a PhD should be is a stopgap for you to fill your time because you can't think of anything else to do. Although this does sound ludicrous, there are many people in academia who did PhDs for this reason, or simply because they were offered an opportunity to do a PhD and felt that they couldn't turn it down.

You should be aware by now that there are lots of people in academia who are 'institutionalised'. By this, I mean that they have never had a job (or a life) outside of academia. This is not something that I can recommend for anyone. Such people are quite peculiar in many ways, and they tend to operate outside of the norms and standards of the rest of society. In particular, they make academia a more difficult place to work, because they tend to accentuate and maintain a lot of the silliness within these institutions. If you are worried about becoming such a person, then I think it would be advisable to take some time outside of academia. For example, if you have never worked but are thinking of doing a PhD, consider taking a year (or two) to do a job before starting a PhD. If it is in a related field it will likely help you greatly. I suggest that it will certainly help keep your feet on the ground, and in touch with some level of reality or normality, during your PhD. You'll also have some experience and connections to go back to once you've finished.

### 1.1.7 Don't rely on becoming an academic

Academia is getting more competitive. Whereas once it was almost certain that you would be employed as an academic after finishing your PhD, this is no longer the case. Indeed, the number of people who get PhDs is greatly increasing the world over, and most are in direct competition for a decreasing amount of academic posts. In 2019, 39% of respondents to a survey said that the most important reason they decided to enrol in a PhD programme was to pursue an academic career (Woolston, 2019). Further, 56% wanted academia as their first choice for a career. Yet in Belgium, for example, the numbers of PhD graduates entering the market nearly doubled (in the first 15 years of the 21st century) while academic positions remained static (Levecque et al., 2017). Universities appear to be exploiting this excess of qualified people to hire them on contracts, reducing the number of full-time positions. Essentially, this means that in most places your time as a PhD student, postdoc and junior lecturer will keep you outside the employers core staff. Joining those on the inside (with well-funded posts) becomes increasingly more difficult. Some have suggested that the world of academia is becoming more like a gang producing drugs (Afonso, 2014).

### 1.1.8 Doing a PhD is really hard work

It's not an MSc with an extra year. It really is much harder and tougher on you mentally and physically. The chances that you'll end up jaded and malcontent with the entire academic system are very high (maybe even obligatory). At times, you are probably not going to be very happy while doing it (there are some exceptions of course), and at these times it will likely have negative impacts on all the good things that you currently have in your life (think family, friends and relationships, as well as free time and fun).

### 1.1.9   Your mental health will likely suffer

Doing a PhD will be the cause of stress not only in your working life but will also impact other areas of your life. Being aware of this at the outset will allow you to alert close members of your non-working life (think partner, family and friends) that they may well need to act as a support network. It is worth tracking your mental health during your PhD to check that you are not getting into difficulties (see Section 1).

### 1.1.10   Any PhD has to be yours

It's entirely useless to do a PhD for someone else. You must take complete ownership in order to do it. In fact, you need to be obsessed with the PhD subject and really want to do nothing else at all. A PhD is an obsession where you can be unwavering and manic in your fascination with a subject. Very few people really understand what a great privilege this is. If you don't feel like a total maniac for learning, a PhD is not for you.

### 1.1.11   Does it matter where you do your PhD?

Not all universities in the world have a PhD program. This means that you cannot do a PhD in some institutions. Larger institutions are more likely to hold PhD programs, have more potential advisors, and may even have funds for bursaries to conduct PhD studies. It is likely that you are going to need to do some work at the university in order to justify any bursary that they may want to give you. Typically this kind of work is the teaching of undergraduates which is a generally good and positive experience in the world of academia.

The country where you study a PhD does not determine the quality of institution (see Figure 1.2). Rather look at the merit of the particular institution that you are interested in. Surveys like that from the Times Higher Education Supplement[3] can help give you an overview.

Depending on where you study might mean that you have different amounts of course time. There are some institutions that will carry on funding you for as long as you want to carry on studying. This is unusual. Most countries now have a restriction on the amount of time that you can spend studying a PhD.

---

[3]https://www.timeshighereducation.com/world-university-rankings/2021/world-ranking

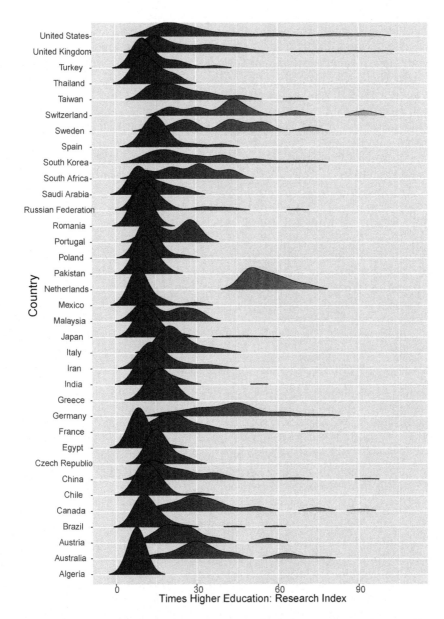

**FIGURE 1.2: It's not the country that determines the quality of your university.** In this plot, you see the Times Higher Education (THE) World Ranking (2021) score for Research plotted by country (with ten universities). Note that most universities in most countries have mean THE Research scores in the lowest quartile. Almost all countries that have institutions in the top quartile have most of the institutions in that country in the bottom quartile. The point here is that you should look at the quality of the institution that you want to attend, rather than assume that all universities from certain countries have high or low research profiles.

It does make a difference where in the world you do your PhD. There are some beautiful places to work. Some places are more expensive to live and therefore survive. If you are a biologist then you might want to consider where your study animals or plants are in order to decide your place to study. But most of all your choice of where to study is likely to be dictated by the presence of an advisor that you want to work with, and who wants you as their PhD student. Advisors are tied to their institutions, and any principal advisor will need to register you at their own institution, meaning that this will be where you spend your desk time. Some advisors might be more flexible than others, but it will be most likely that you will need to move to the institution where your advisor is based.

## 1.1.12   What to look for in an advisor

Throughout this book, I will call the person who is the academic in charge of your PhD studies your 'advisor'. I will not call them your supervisor. The reason for this is because I think it is much better for you to think of this person as giving you advice and not the person who tells you what to do. This comes back to the idea that the PhD is yours. It is not your advisor's PhD. They will not write it. They will not do all the work that's required for it. That will be yours alone.

Just as I've attempted to do in this book, that person will give you advice. It will be up to you whether or not you follow that advice. Many of my students have not followed my advice. For some, this has been the right thing to do. For others, it has not. No one can know what the best thing to do is in different situations, because they will almost always be unique. You will be the expert on your project. That means that as the person closest to your project, you are the person who is best placed to make a decision. But your advisor can, and should, help you to make those decisions.

Your advisor can help because your advisor has experience of doing research themselves. They hopefully have some experience of advising other students. And they should have some experience of having done research in your chosen study area before, including having published papers in appropriate journals.

The advisor that you're looking for should be offering or doing projects in which you are interested. There may only be a few people in the world that are doing this type of research, and you may end up having to choose between them. Hence, while you're reading about how to pick a good advisor, your shortlist may only be of two or three people in the world. The project and subject area are going to be strong areas driving your choices moving forward.

One of the best ways of picking a good advisor is by talking to current or past students. You should ask them how happy they are with their advisor and whether they are receiving the support that they need to do their projects. A good advisor should have an open door so that you can talk to them whenever you need to. Thus, once you have talked to the students in their group, you should also make an appointment to talk to them about your project. It might not be one meeting, there may be several. The more contact you have with them, the more likely you are to be sure that this is the person whose help that you want when conducting research over the next few years.

Good advisors tend to have good strong research groups that work well together. You can look through the publications of a research group to see how well they work together. You should see the names of many of the graduate students appearing on publications demonstrating the level of collaborative engagement within the research group.

A good advisor should have a good network. Both during your PhD studies and after them, you will rely primarily on the network of your advisor as a basis to building your own network. For example, you will go to meetings that they go to, you will be introduced to people that they know, and you may have opportunities to collaborate with people within their networks (see Adams, 2012). Doing a PhD with an advisor that has no network and that does not link up with other academics (or worse is disliked by other academics) might mean that your own work is not recognised, or even sidelined. You can easily check on the network of a potential advisor by looking at the collaborators in their publications, or on their website.

Having said this, the best academics also tend to be very busy, and that may mean that they do not make good advisors. The balance is definitely very tricky, and one of the best ways to make sure that they have enough time for all of their students is to limit the number of students they have. Clearly, this works against prospective students.

### 1.1.13   What to avoid in an advisor

Remember that relationships occur on a one-to-one basis, and it may be that not every advisor gets on with every student. There are some advisors who have many problems with many students, and these are best avoided no matter how high their profile is (see Malaga-Trillo and Gerlach, 2004; Abbott, 2004; but also see Vences and Zardoya, 2004, for alternative viewpoints). There are other advisors who are so busy that they are never able to meet with their students, and pass this task on to somebody else in their group. Again no matter how high profile these people are if they can't advise their own students they are best avoided.

At the end of the day, there is no such thing as the perfect advisor. You will likely have to make a trade-off on the above points with someone who has a project that they are willing to advise you on, and your chosen area of research.

## 1.1.14   What makes a good project?

After having made lots of statements about how a project should be your own project the following advice might seem a bit strange. Clearly, the work that you're going to be doing will need to fit into the group of the advisor that you've chosen.

The most obvious point to make here is that a project should be something that you want to do and that you're prepared to dedicate yourself to for at least three years.

Projects that are cross-disciplinary are likely to lead to more opportunities in more disciplines. You may want to make sure that you do not narrow your focus entirely into a single technique or method. If you view your PhD as a training experience, then it is better not to be doing exactly the same techniques that you have mastered previously, for example during your MSc. It is worth bearing in mind that academia has trends and fashions and that becoming an expert in a brand new technique could land you a job and career. It could equally mean that your entire knowledge of a subject area is replaced by a better approach within one or two years.

## 1.1.15   Funding your PhD studies

Funding may be one of the biggest obstacles that you come across when wanting to do a PhD. It's important that you consider the funding of your PhD in two separate ways.

### 1.1.15.1   The funding that you need to live

This money is typically called a bursary or stipend depending on where you are in the world. It seems to me that many countries offer bursaries that are not actually enough to live on (e.g. buy food, rent a house, etc.). If you're offered a PhD bursary, especially if it's in another country, then do check on the living cost at that location and whether or not the bursary that you are being offered will be sufficient money in order for you to survive. Also, be very careful to check how long the bursary is for. For example, at my university a PhD bursary is given only for three years. After this three-year funding period, you will be expected to have submitted your PhD for examination. If you need further funding to live while your PhD is being examined, and before you graduate then you will need to get a job. There is no more money!

It's a really good idea to talk to people who have already done a PhD to find out more about the different available bursary opportunities, and how people manage with and without them. There is no one single model, and the more people you talk to the more likely you are to discover one that might suit you.

Of course, conferring with other people can include reading the accounts of many different people who share their experiences on social media. There are blogs and Twitter accounts dedicated to this stuff. Try and be as open-minded as you can.

### 1.1.15.2 The cost associated with your studies

The money associated with the project that you are doing also needs to come from somewhere. Depending on where in the world you can do these studies there will be different expectations of where this money comes from. This is a very important topic for you and your potential advisor to discuss. During the proposal phase of your PhD you'll be expected to put together a budget to determine the total costs of your studies. Ideally, you or your advisor should have a ballpark figure of how much this is going to cost and have some idea about where the money is going to come from.

Again if you are moving to a new location to do a PhD then beware that costs are not the same in different countries. For example, you might be used to paying very little money for accommodation and travelling in the field, essentially meaning that you can do as much fieldwork as you like. But in some countries fieldwork will make up such a substantial proportion of the funding, such that you may be limited in the amount of field work that you can afford to do.

It is possible that your potential advisor will tell you that they don't have any funding for a project that you may want to do. This is both a curse and an opportunity. There are many opportunities to raise money to fund PhD projects. Raising the funding yourself is one way of making sure it really remains your project. But the availability of this kind of funding does vary from country to country and you may want to embark on the fundraising side of your PhD before you register. Especially if you have a limited amount of time once you have registered in order to complete your PhD studies. Some projects are much cheaper than others. For example, if your intention is to sequence genomes of every animal or plant that you work with, the finance required will likely be far greater than your bursary. If however your intention is to do a PhD that models distributions of populations of animals and plants, then the entire costs may only run to a computer remote access to a computing cluster and data storage.

Funding, and the availability of funding, may well be the tipping point that makes you decide where you will do your PhD studies, and with whom. If you haven't already got the message (see above) then these are important decisions and you may be better off postponing the start of your PhD in order to raise money, and do the project you want, in the place you want, and with the person you want to advise you.

# 2

## Reconciling Yourself to Doing Things You've Been Avoiding

Some 30 odd years ago, as an undergraduate, I was told that I would have to use computers in one of my second-year courses. The thought filled me with dread and I remember putting off going into the computer room for as long as I could. It was a steep learning curve, and I struggled. As my only previous experience had been with a typewriter, each time the cursor got to the end of the screen I hit the, 'carriage-return button'. And each machine had an odd white device attached called a mouse, and I had no idea what that did. All this seems quite preposterous now, but imagine embarking on a PhD in biological sciences without reconciling yourself to using a PC. This is practically where we are now with R (R Core Team, 2021, Figure 2.1), and a wealth of other alternative and emerging programming platforms.

**FIGURE 2.1: The ubiquitous R logo.** R (and the invaluable GUI RStudio) has become the go-to platform for many statistics and figures in biological sciences

If you are not already aware of the prominence that R now has in conducting statistical analyses in biological sciences, then you've missed out on a great deal. The real point here is not that you can use R, but that R represents a platform on which analyses can be easily repeated, and by saving and making the code available together with your data, you are making your data analyses reproducible. We will see elsewhere the importance of reproducibility in sciences (Baker, 2016), and the critical role this has in transparency and, therefore, best practice.

What is perhaps most remarkable about learning R is the freedom it gives you away from large amounts of commercial software. As a basic programming

DOI: 10.1201/9781003212560-3

language it can be used for statistical analysis, drawing figures, conducting geographic information system (GIS) level work, and even as a word processor on which to write your thesis (and indeed this book Xie, 2016).

If you already know all of the above, then you won't face any major barriers, and you will fit into the new and more transparent world of science with great ease. If the thought of using R fills you with dread, in the same way that using a computer did for me some 30 years ago, then you should reconcile yourself now before you go any further. Learning R will be a steep learning curve, but it will be made easier by the large amount of excellent video tutorials and online courses that are available.

## 2.1   And learning lots of other stuff...

The same principle that I've written about above can be extended to all sorts of other aspects of your PhD project. You will have to learn new skills, and some of these will be things that you have previously resisted having to do. Allow yourself to be open to all aspects of learning, and all skills. The act of learning opens new pathways in your brain, and using this 'muscle' will facilitate future learning exercises.

# 3

## The Scientific Project and Scientific Living

Throughout the book, I make reference to the 'scientific project' as a greater plan to which our studies in biological sciences is simply a part. I consider the scientific project as a path that humanity should continue to travel in order to reach a higher and more just way of living. Our societies have evolved from places where superstition and belief were the basis for societal rules that favoured the privileged classes. Science is a process in which we reach conclusions based on egalitarian reasoning and evidence, something that we can aspire for the larger community around us. Already, most societies are making the majority of their collective evidence-based decisions through the guidance of science. Therefore, if we consider our knowledge-based societies to be democratic, then all citizens of that society should understand the processes under which the foundation of that knowledge is acquired (Sagan, 2011). Yet globally, public skepticism in science has been rising, and has reached as high as 35%. To consolidate this growing position of science in society, and the decline in public acceptance, scientists will need to become more open, communicative and seen to be conducting their own profession in a way that reflects these ideals. There is evidence that increasing communication from scientists to the public results in a decrease of skepticism (see here[1]).

Imagine a world in which all of the principal decisions are made as a result of advances in science, but vanishingly small numbers of people actually understand how science works. This is not a world that we would want, but as Carl Sagan (Sagan, 2011) pointed out, it's one that could all too easily happen. Therefore, to prevent the workings of science falling into the hands of a minority, we have a duty to communicate our science to the majority.

Ultimately, what we want from the scientific project is to draw everyone in. That is not to say that there is no place in the world for other subject areas. One of the greatest misadventures in my own education was the indoctrination that I received that there were two streams in education, one for STEM and one for arts. I hope that you managed to escape this narrow minded and factually incorrect view, because as you will discover there are high levels of creativity demanded within the scientific framework, and those who have more talent in communicating their science will likely be more successful in their

---

[1] https://multimedia.3m.com/mws/media/1898512O/3m-sosi-2020-pandemic-pulse-global-report-pdf.pdf

DOI: 10.1201/9781003212560-4

careers as a result. Moving forward with the scientific project does not mean that everyone must become a scientist, but instead that the process by which scientists establish knowledge is understood. Indeed, as we shall see (below), sceptical thinking, as exemplified in science, is a healthy way for everyone in society to avoid time wasting and potentially being deceived or conned. It is also vital that the scientific process be transparent in a way that results can be understood and therefore respected.

Here I provide an example of the way in which there is a need for efficient communication between biological sciences and civic society.

## 3.1   Example: Invasive species

Alien species are those that have been moved by human agents across biogeographic barriers into parts of the world where they did not evolve. Once these aliens arrive in their new environment, some reproduce and become established, and after a lag, a fraction of these spread and become invasive, often with substantial impacts on the environment and the socio-economic efforts of people in those invaded areas. Invasive species disproportionately affect poorer people in our society, often completely disassociated with the pathways responsible for the introduction and spread. Removing invasive species requires a considerable amount of effort, and unless this is supplied by taxpayers, via governmental policy, falls on those directly impacted. Invasion science, the expanded transdisciplinary area stemming from invasion biology, is required to engage with all stakeholders in invasions, and most importantly to prevent new invasions. Because introductions are made by humans there is a critical need to communicate with those stakeholders along introduction pathways.

For existing invasions, social scientists and economists are needed to determine their impacts on people in invaded areas and determine how to best alleviate the invasion or mitigate its effects without it moving further and impacting more communities. Biological scientists need to assess the impact on the environment, and investigate ways in which the species can be removed, their spread curtailed, or their impacts neutralised. Importantly, the results need to be discussed with all stakeholders to achieve a desirable outcome for those impacted most. Invasions at larger scales will normally require governmental resources to ensure that communities can continue to function, especially to best deliver their services to the society. Governments require evidence-based decisions that are reached from well-researched studies conducted in the field. Effective measures may necessitate large sums of money, and governments require that this process be transparent so that the taxpayers' money they are investing is spent efficiently.

We know enough about invasive species to be aware that there are greater threats to society from species that have not yet arrived than those already present. This means that all societies need to guard against the threat of potential future invasions, and still maintain the trade that brings needed goods into their areas. Most societies have no checks in place, but the immediate need to prevent new invasions, and halt the effects of existing ones, is recognised on a global scale. Biologists have an important role to play as key agents in informing society on the impacts (especially when they disproportionately impact those least privileged in society), pathways, removal and prevention of invasive species. Failure to effectively communicate with any of the stakeholders along the invasion pathway will result in more propagules arriving, and with them the increased chance of more species becoming established and eventually invasive.

In the case of biological invasions, communication needs first to flow from the various stakeholders to direct the research needed. The solutions required are context specific, and the lessons learned need to be communicated back to the stakeholders, including governmental policy makers, in the form they require to action the findings. Ineffective communication from scientists to the stakeholders may result in solutions being ignored or misinterpreted. Documentation disseminated to the global scientific community, through scientific publications, is needed in case it may help the response of other societies with similar problems.

## 3.2 Should society lead the sciences or sciences lead society?

Of course, the answer is simply that there must be feedback between science and society. As science advances, society learns of these advances and views are advanced. To become incorporated as a societal norm takes time. Over time, views from science that appear radical to society become mainstream and accepted. Think of the views on the earth being round, or more recently, the general acceptance that the universe started with the 'big bang'. The counterculture, also known as post-truth, faction in society that espouses views accepted by the mainstream and proved by science beyond reasonable doubt is not a novel phenomenon. Indeed, in the scientific project, we need to accept that these views will persist and empower the general populace to understand why and how they can be shown to be false. Hence, your view that the earth is not flat is bolstered if you can provide three ways to demonstrate this as fact (pick the ones you like from seven listed here[2] or here[3]). Otherwise, your assertion that it is not flat will appear as easily denied as a 'flat-earther's' assertion that it is. Thus, the power of the scientific argument in society is that it is not simply contradiction or gainsay, but that there is evidence.

The scientific project is currently not the open virtuous eutopia that we would like. Indeed, as scientists are humans and part of the current society, they are prone to the same trappings of wielding power and influence in order to get ahead. As you get towards the end of this book, you will see me calling out increasing parts of the scientific project that are currently riddled with bias, thereby weakening the entire process and disenfranchising the most vulnerable. Reforming science is no easy task. Reforms are needed throughout the structures that fund, employ and disseminate scientific research. Sweeping this all away will take a concerted effort but might well be easier in science than in other sectors of society. Nearly all scientists recognise that science needs to be objective and impartial in order to succeed. Their objectivity makes them an ideal group to kickstart a fairer, more transparent and open system. As you will see, there has already been a move to make science more open, and by embracing this transparency the next generation of scientists will reap the benefits, and be on a sounder footing to improve a better scientific society.

## 3.3   Carl Sagan's "Baloney detection kit"

### or **"How to know when what you smell is bullshit"**

In his book, the Demon-Haunted World, Carl Sagan (2011) provides his take on how science provides scientists with what he terms 'Baloney Detection', commonly referred to today as 'bullshit'. These are well worth recording here as they provide a guiding light for all scientists whether they be novices, emerging or established. They are also easily taught and the foundation of what we should spread in the Scientific Project.

As humans, we are all susceptible to bullshit, hoaxes, charlatans and contrived deceptions as they are often more appealing than the truth. However, science is about truth-seeking, and by applying a set of simple rules, Sagan argues, we can avoid being misled or distracted by ideas that may waste our time and energy by using 'skeptical thinking'. These rules are useful in your scientific life when considering hypotheses or when building arguments to explain your results. However, as Sagan argues, these simple rules are also useful in your

---

[2]http://crosstalk.cell.com/blog/seven-ways-to-prove-earth-is-round
[3]https://www.unlv.edu/news/release/round-earth-clues-how-science-proves-our-home-globe

day-to-day life, and as a simplified example of the scientific method, we can pass them onto all of our fellow citizens.

1. You should always seek independent confirmation of results.
2. Seek out different opinions on the results and consider all possibilities.
3. Don't believe facts simply because they are spoken by an authority on the subject. We know that humans are prone to confirmation bias, and often exhibit motivated reasoning, especially if they have some vested interest. Even the most experienced have made mistakes in the past and will again in the future. My preferred way of thinking of this rule is to acknowledge that experts are made of ex- (a has been) and -spurt (a drip under pressure).
4. Use a hypothetical deductive framework. Instead of accepting the first (hypothesis) explanation, think of all the possible reasons why a result might have occurred and then consider how you could (or perhaps you already have in your experiment) disproved each one. You may find that an untested hypothesis to explain the results represents a major caveat that could be tested, or can't yet be tested. Either way, you need to remain open to the possibility that this explains your results. This leads neatly onto:
5. Don't get too involved with keeping your chosen hypothesis opening you to potential confirmation bias. Remain critical to the possibility that it might be wrong, and provide ideas on how to substantiate or disprove competing hypotheses. If you don't, others will.
6. If at all possible, use a quantitative method to test your hypothesis. You should already be well on the way to this one inside your PhD, but it is worth bearing in mind, especially when it comes to considering testing other competing hypotheses. How could they be quantitatively tested?
7. Arguments consist of a chain of statements that follow one from the other. This chain is only as strong as its weakest link, and the strength of each link in the chain (your argumentative statements) must be critically considered by you.
8. Occam's Razor dictates that given two hypotheses that equally explain the results, the simpler one is likely more plausible. Or put another way, the more links you have in your chain, the more likely that one of them is not as strong as you would like it to be.
9. Your hypothesis (or argument) must always be falsifiable. That is, that your test should be able to find that it is incorrect. If your test cannot show that your hypothesis is incorrect, then it isn't worth much.

Sagan provides plenty of real-world examples of each of these arguments, and I'd suggest that you consult his book (Sagan, 2011) if you are interested in following these up. Similar approaches, but written in different styles, are provided by Richard Dawkins in his (2004) book: A Devil's Chaplain, and Julia Galef in her (2021) book: The Scout Mindset.

## 3.4 Live your life scientifically

**"Science is more than a body of knowledge, it is a way of thinking"** Carl Sagan (2011)

I would encourage you not to stop your objective and evidence-based thinking when you leave the office or laboratory. By applying the rules of science to other sectors of society, you will find that there are many societal norms that could benefit from similar reform. Certainly, a lot of the technology developed through scientific inquiry can be used to help society reform, but scientific thinking will help everyone. As well as bringing the world to scientific thinking, as scientists we should take scientific thinking to the world.

## 3.5 Citizen science

For a long time, I struggled with the term 'citizen science' as I reasoned that many such ventures did not involve citizens doing science. To me, most seemed focused on outsourcing large tasks to willing public participants: i.e. having citizens do some grunt work. Collecting data or doing a fraction of the analysis is not science. However, in light of the scientific project, I have changed my mind. All of these projects offer citizens a way to participate in scientific projects in a way that was previously unavailable to them. They promote direct communication between scientists and citizens. In this way, they serve to promote the understanding of science to the general public.

One aspect of this will be to bring more of the population into conducting scientific studies. This can be at the level of funding scientific studies. Crowd-sourced scientific studies promote communication between the scientists and the funders to a level that is meaningful to both groups, such that they make up a large scale collaborative research interaction (Uhlmann et al., 2019). Citizen science contributions mean that the collection of data can be shared on a much broader scale than would be traditionally possible. Of particular interest in the biological sciences is the identification and collection of locality data for biological species (on platforms such as iNaturalist). But increasing numbers of projects are demonstrating novel ways for the public to increasingly participate meaningfully in scientific endeavour (Bonney et al., 2014; Silvertown, 2009). Science is a joyful and spiritual experience for most of us that participate in it. There's no reason not to spread that joy as widely as possible.

# 4

## Keeping Track of Your Mental Health

Stress is a natural part of life and many people are at their most productive when they are under some degree of pressure, such as a deadline. Although deadlines don't work for everyone.

Douglas Adams famously claimed:

"I love deadlines. I like the whooshing sound they make as they fly by."

Problems arise when we become overwhelmed by stress and are unable to fully respond. When this occurs, productivity can drop off and survival responses can be triggered as if responding to an actual physical attack. These responses include fight, flight or freeze responses. Anxiety and panic can be triggered. In this state, additional demands on your time may also push your life off balance so that you start to neglect your personal wellbeing which can negatively impact on relationships, exercise regime, or even nutrition and personal hygiene. Some people can find that the additional stress can cause physical symptoms that may even need medical treatment. Your sense of competence and mastery can be negatively impacted such that you may even suffer from feelings of inadequacy or impostor syndrome (see Part II).

Although there are not many studies on mental health for PhD students, those that exist (as well as surveys: Nature 2019[1]) all suggest that there is a significant toll, which is proportionately higher than for others in society (Levecque et al., 2017). Whatever your prior experience of stress in a working environment, academia is known to be particularly stressful, and as a PhD student, you are likely to absorb a significant amount of this stress into your own life (Stubb et al., 2011).

---

[1] https://www.nature.com/articles/d41586-019-03459-7

DOI: 10.1201/9781003212560-5

The General Health Questionnaire (see GHQ-12 in Table 4.1) is an instrument used to measure psychological distress. It is quick, reliable and simple to score, so you can use it at any time during your PhD studies as an indicator of whether you need to reach out to personal, occupational or professional support networks.

Right now, I suggest you complete the GHQ-12 (Table 4.1) and record your answers as a baseline. Keep the scores somewhere safe. During the course of your PhD, if you feel that your scores may have changed, take the test again and compare them with your baseline scores. Although there are no hard rules, if three or more of your scores have moved by two or more points it could be worth discussing with your support network to help you decide whether or not to seek professional help.

**TABLE 4.1: A General Health Questionnaire with 12 questions (GHQ-12) that you can use to keep track of your mental health.** Answer the questions in this table now, and compare your answers to answers in your future when you are feeling more under strain. Write down any explanations you have for your choice of answers.

| General Health Questionnaire: Have you recently... | 0 | 1 | 2 | 3 |
|---|---|---|---|---|
| been feeling reasonably happy, all things considered? | Better than usual | Same as usual | Less than usual | Much less than usual |
| lost much sleep over worry? | Not at all | No more than usual | More than usual | Much more than usual |
| been feeling unhappy and depressed? | Not at all | No more than usual | More than usual | Much more than usual |
| felt you couldn't overcome your difficulties? | Not at all | No more than usual | More than usual | Much more than usual |
| felt under constant strain? | Not at all | No more than usual | More than usual | Much more than usual |
| felt capable of making decisions about things? | Better than usual | Same as usual | Less than usual | Much less than usual |
| been able to face up to your problems? | Better than usual | Same as usual | Less than usual | Much less than usual |

| General Health Questionnaire: Have you recently... | 0 | 1 | 2 | 3 |
|---|---|---|---|---|
| been thinking of yourself as a worthless person? | Not at all | No more than usual | More than usual | Much more than usual |
| been losing confidence in yourself? | Not at all | No more than usual | More than usual | Much more than usual |
| been able to enjoy your normal day-to-day activities? | Better than usual | Same as usual | Less than usual | Much less than usual |
| been able to concentrate on whatever you are doing? | Better than usual | Same as usual | Less than usual | Much less than usual |
| felt that you are playing a useful part in things? | Better than usual | Same as usual | Less than usual | Much less than usual |

Even if you don't feel you need the support of your institution now, it is worth finding out how they can support your mental health in the future if needed. Although there has been some stigma attached to difficulties with mental health in the past, most institutions accept that pressures are mounting on postgraduate students and that they may require support. Most institutions have experienced councillors available to support you if needed. Importantly, you should realise that none of these symptoms are unusual and that there is a high probability that many of your colleagues may also be struggling. Knowing that your problems are shared and reaching out to support networks early is an excellent way to prevent them from escalating beyond your control.

A study into the mental health of PhD students in Belgium exemplifies the kinds of difficulties that they face when compared with other similar groups (Levecque et al., 2017; see Figure 4.1).

No matter how well you think of your own abilities to cope with mental health issues, doing a PhD will cause you additional stress and can trigger maladaptive coping mechanisms. Learning how to cope with additional stress early in your career can be beneficial for future personal development.

Academia is recognised as a particularly stressful environment; you will likely take on some of this environmental stress in addition to any stress associated with your studies. Additional stressors come from home and family situations. Your best means of coping will be to try and develop a support network and

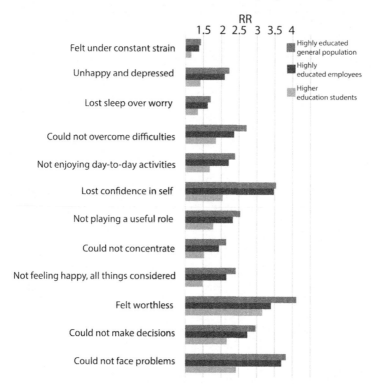

**FIGURE 4.1: A comparison of the mental health of PhD students** (data from Levecque et al., 2017) with highly educated general population, highly educated employees and higher education students using the General Health Questionnaire. The risk ratio (RR: adjusted for age and gender) in PhD students in Flanders, Belgium, is consistently higher (approaching 1) when compared to any of the other surveyed groups. The closer RR is to one, the higher the levels of strain.

to understand where and with whom you can discuss any difficulties as they arise. Knowing who this is and how and when to approach them will put you in a stronger position if you need them in future.

## 4.1   Being physically active improves mental wellbeing

The positive relationship between the amount of physical activity and higher mental wellbeing is well established (e.g. Grasdalsmoen et al., 2020; Gerber et al., 2014), but the kind of exercise required to achieve this improved result is varied. For example, a study of college students showed that an hour of Thai Chi twice a week for three months was enough to have significantly improved physical and mental health scores (Wang et al., 2004). Increased levels of physical activity were also found to be associated with improved sleep quality in another study (Ghrouz et al., 2019; Gerber et al., 2014). There are plenty of studies out there that suggest there are multiple benefits from physical activity.

If you don't do it already, then when you start your PhD start regular exercise. Getting fit was the best thing I ever did to improve my work. The first thing I found was that I could sit and concentrate on work for longer. Before I got fit, I was constantly distracted and needed to take lots of breaks. Now I find that my concentration is much better and that as my fitness stamina increases, so too does my ability to concentrate (see Williams 2021).

There are other ways to achieve these benefits, including mindfulness or meditation.

## 4.2   Time to think

I used to think that the time I spent exercising was unproductive with respect to work. Now I have begun to realise that this represents some of my best thinking time. I use the time I spend trail running to turn ideas over in my head and especially to think through the logic in arguments and potential flaws in experimental design. This isn't to say that I don't need to put time in at the computer writing it all down, or talking through ideas with colleagues. But I find the meditative time I spend exercising to be especially productive.

## 4.3   Balancing work with life

Having already told you that doing a PhD might unbalance you mentally, regular exercise is a really good way to make sure that you retain that balance with life (see Hotaling, 2018). By life, I'm referring to everything away from your academic work. This might include family, friends, sport, hobbies (even fishing). If you don't have friends outside your academic life, then it would be a good idea to find some so that you can keep in contact with the 'real world'.

If you already exercise, then you likely know how important this is. But if you don't then, my best advice (above anything else that you might read in this book) is to start as soon as you can. Do whatever is right for you. If you don't already know, then try out different clubs or activities at your institution. There are normally lots to choose from. Make sure that you schedule regular time for your activity, even if that means that you run in the mornings before field work starts, or you spend your evenings doing Tai-Chi on the beach after a day at sea doing research on dolphins. Take your exercise with you wherever you go for your studies. Take it as seriously as you do your studies and you'll find that both will benefit.

Once you have a healthy body, you'll find that there are all sorts of things that go along with this. The desire to fill it with healthy food resulted, and this later tied in with my interest in the gut microbiome. I now have many fermented foods in my kitchen (much to the horror of my kids), and the improvement of my digestive system. I could go on, but I'll spare you. Instead, I just want to emphasise that I consider that not only will your quality of life improve, but so will your work if you include regular exercise as part of your PhD regimen.

# Part II

# The Nuts and Bolts of Writing for the Biological Sciences

DOI: 10.1201/9781003212560-6

# 5

## How to Get Started with Writing

**Before you start writing**

The writing part of the PhD is often the most difficult for many students. I think that there are several reasons for this. Writing well comes with practice, and most PhD students have had little experience with academic writing – even if they have already completed an MSc. The same could be said of most early-career scientists (even after their PhDs) because practice is measured in completed pieces, it usually takes many years rather than weeks or months. You shouldn't expect that scientific writing will come automatically or immediately. In order to get this practice, you need to get started, and (again) many students find starting to write challenging. This guide was written to help you get started.

However, before you start writing, it's a good idea to understand some of the ways in which academic writing differs from other writing styles. Good planning is the key here, along with knowing exactly what it is that you'll be writing about. That sounds facile, but at the foundation of each scientific study lies the hypothesis, which is central to being able to build the rest of the chapter.

In this part of the book, I start with an explanation of the hypothesis. As you will see in Part III, the hypothesis is central to all of the writing in each chapter or paper, and you need to have a very good idea of how to formulate your research question into a hypothesis before you begin. Then I build on helping you think through lots of different aspects of writing in science: style, citations, writing concisely, scientific names, and lots more. Finally, I introduce you to the idea of using a formula to structure your writing.

Before you start writing in earnest, it'd be good to read this part of the book and make sure that you are familiar with the ideas herein. Later, you may need to refer back to this section once you've started writing.

DOI: 10.1201/9781003212560-7

## 5.1   How to write a hypothesis

When you read about the formula of writing a chapter (see the end of Part II), you will see that everything in your chapter or paper revolves around the hypothesis or questions that you ask, normally at the end of your introduction (Figure 5.1). In this section, we will take a look at how to write a hypothesis. This is a sticking point for many students. We are used to using and writing questions and statements in day-to-day communications, as well as reading popular media. But hypotheses (the plural of hypothesis) only rarely float across our desks. So how do we write one, and how do we know if our hypothesis is good?

In addition to this section, there is some good information out there on the web, and it's worth looking at this too: (e.g. Wikihow[1], Wikipedia[2], etc.). There's also some less good stuff out there, so read critically.

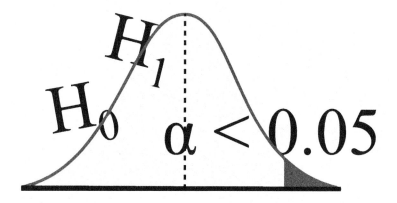

**FIGURE 5.1: Generating a good hypothesis for your PhD chapter isn't easy.** A good hypothesis will be invaluable in helping you write the chapter. It is important to start working with hypotheses.

### 5.1.1   What is a hypothesis?

A hypothesis is a statement of your research intent. It tells the reader (because just like all of your other written work, it is intended for an audience who will read it), what you plan to do in your research. But there's a little more

---

[1]https://www.wikihow.com/Write-a-Hypothesis
[2]https://en.wikipedia.org/wiki/Hypothesis

to it than this. The hypothesis becomes a part of the scientific method if it is testable and (importantly) falsifiable, as well as being informed from previously published work on the subject.

Your hypothesis must be informed by the literature, which is why you spent so much time and effort crafting your introduction to inform your reader of the same. This is also why your hypothesis usually comes at the end of your introduction because you spend all of the introduction telling your reader about it. There's not much point in writing more of the introduction after the hypothesis because once your reader has read that, they are ready to learn about how you went about testing it (in the Materials and Methods). The other important point to make is that the literature should dictate how you write your hypothesis, and the variables that you include. If, for example, you know that temperature is the most important variable, but all of the literature suggests that it is oxygen, you can't ignore oxygen and you should also frame your hypothesis using this variable (you can have more than one hypothesis after all!). In this case, you will also need to provide a sufficient introduction to temperature as a variable to justify its inclusion in your hypothesis. Perversely, your aim is not to prove that your idea is right but to show that the hypothesis can be refuted.

We try to write a hypothesis that is falsifiable: i.e. you can prove (often usually using statistical tests) that it is not correct (or at least show that the likelihood that it is correct is very low). That's why it is conventional to provide the 'null hypothesis', that is the version of the statement suggesting that there is no relationship between the variables you have proposed to measure. The convention is to label this 'null hypothesis' $H_0$, while the 'alternative hypothesis' (the one that says your variables are related as you suggested) is written as $H_1$. When you formulate your hypothesis, it is traditional to write your alternative hypothesis to indicate the directionality of your tested variables. This way, the reader can simply imply that the null hypothesis is when there is no relationship, but this will need to be stated if the null hypothesis is more complex.

Karl Popper (2005) was the philosopher who proposed that without being able to refute or falsify a scientific problem, it ceases to be scientific. This is the reason for our null hypothesis. If the null is not available as a possible outcome, then logically, there is no science.

---

Karl Popper (2005):
**"...it must be possible for an empirical scientific system to be refuted by experience".**

---

Most importantly, your hypothesis must come first before you do the experiment or study. Hence the reason why this chapter comes at the start of Part II. Setting the hypothesis after the work is already done is fraudulent, and goes against the scientific method (also known as HARKing). Obviously, it isn't fair to pose the hypothesis once you already know the answer. This is why there is so much emphasis put on formulating your hypothesis during your research proposal. Getting it right will determine what you do and how you test it. If you think of an extra hypothesis that would be really useful to test once you've already done your study, you can conduct a *post hoc* test, but this should be clearly labelled and have more stringent levels of statistical assessment.

Writing a hypothesis isn't easy, but it is essential, and once you've understood what to do, most of the rest of what you are writing for should make sense.

### 5.1.2   What a hypothesis isn't

It is not a question and so should never have a question mark after it.

It isn't really a simple prediction: if this then that. You will see on the internet that hypotheses are explained in this simple predictive framework. I say that a hypothesis is not a simple prediction because it lacks the mechanistic and scholarly aspect of a good hypothesis, which is what we want to achieve.

### 5.1.3   A formulaic way to start writing your hypothesis

**"If. then. because."**

Later in the book, I will emphasise that you must have introduced all the variables that you plan to use to test your hypothesis in your introduction. This usually comes in the second paragraph of your introduction, where you emphasise the utility of the dependent variable/s (what you are planning to measure) and your independent variable (what you will manipulate). Both of these variables should then feature in your hypothesis. Next, by paragraph four, you will have identified the problem that you are interested in tackling. In addition, your introduction will provide all of the pertinent literature that has relevance to this hypothesis, giving the all important context.

A simple way to consider making your hypothesis is to adopt an **"If. then. because."** construction where you add in your problem statement using your independent variable after 'if', and your prediction using your dependent variable after 'then', and finally the expected mechanism after 'because'. Using our example above with the "If. then. because." construction, we would say: "If environmental temperatures in which our study organism develops are increased then development rate is faster because they follow the classic metabolism of ectotherms". Both independent variable (temperature) and dependent variable (development rate) are present in this hypothesis, and the predicted relationship between them is clear. In addition, the causal mechanism is stated. I say that this is a formulaic way to start writing your hypothesis because it usually ends up as an inelegant statement, which can be better refined for a reader. A citation for your stated mechanism might also help clarify exactly where the justification for this comes from.

Mechanisms (or causal explanations) fall into three main areas: endogenous, exogenous and evolutionary (Allen and Baker, 2017).

### 5.1.3.1 Endogenous causal explanations

Endogenous causal explanations focus on the mechanisms happening inside an organism, such as physiological processes, hormones, reproductive state, etc.

### 5.1.3.2 Exogenous causal explanations

Exogenous causal explanations concern mechanisms that are outside the body of individuals. Common exogenous mechanisms are climatic factors (temperature, humidity, precipitation, etc.) or may relate to the availability of food, predators or mates.

### 5.1.3.3 Evolutionary causal explanations

These mechanisms have evolved through time and often relate to exogenous mechanisms triggering endogenous processes over multiple generations.

Note then that the above mechanisms are not mutually exclusive in their nature, and it may be useful to combine different approaches within biology to ask hypotheses across all of these levels. Mechanisms in biological sciences are rarely simple or act on multiple organismal levels, so designing a controlled experiment in order to test a specific mechanism thoroughly can be very demanding. In other words, can you be sure that the cause is really responsible for the effect that you are measuring?

A good hypothesis will often take an existing hypothesis further to try to better refine knowledge on a subject. Hence, it is perfectly acceptable to state that you are building on existing hypotheses (and giving the appropriate statement) when making your own.

### 5.1.4   Teleological versus causal hypotheses

A teleological argument refers to the reason or a purpose of a particular process. For example, you may measure vertical migration of water fleas and suggest that diurnal migrations are made because the water fleas want to avoid predation. This is a teleological hypothesis because you are suggesting that the reason behind a process is the desire by water fleas to avoid predators. Although a reduction of predation may be a consequence of vertical migration in water fleas, each water flea does not think about predation and then start it's upward movement as a result. A common mistake made in biology is to apply teleological arguments to processes that have no purpose or reason. Evolution is often mistakenly suggested to have a purpose (e.g. to evolve to a more advanced state), but in fact, evolution is not a goal-orientated process. There is no end-point to evolution, and evolution did not start in order to meet some predetermined form or function. On the other hand, a causal hypothesis focuses on the factors about A that cause B.

You should have realised that biologists are principally interested in causal hypotheses because most mechanisms that are studied in biological sciences have no predetermined goal. If you are a behavioural ecologist, then you will need to be particularly aware of these two types of hypotheses, and when teleological explanations may be appropriate: many types of behaviour are goal orientated.

### 5.1.5   How to evaluate your hypothesis

Once you've written your hypothesis, how do you decide whether or not it is good? To do this, you might think that you need plenty of experience (and yes, that does help). But really, you just need to look for the elements that are discussed above. So once you've written your hypothesis, try to objectively answer the questions below:

- Is there a clear prediction (if. then. statement)?
- Does the prediction use independent and dependent variables correctly?
- Is the mechanism supported by the literature?
- Is the hypothesis testable/falsifiable?
- Does the hypothesis use concise wording and precise terminology?

If your hypothesis meets all of the criteria above, then you've done a good job!

Probably one of the hardest issues that you will face in biological sciences is to determine whether your dependent variable is reliant only on your independent variable of choice. For example, a lot of variance in biological sciences will relate to the climate (especially with global change studies), but if your independent variable is temperature, this means that you will need to keep all other climatic variables the same. That is, if temperature is your independent variable, it is the only variable that can change in your experiment. This type of experiment is challenging as temperature often affects other variables (especially that they may vary in an unpredictable way). As soon as you have more than one independent variable, you can no longer test your dependent variable because you don't know which independent variable it is reacting to. Isolating variables is notoriously difficult, especially when we move from the laboratory to the field. You will need to think very carefully about what variables other than those of interest are potentially impacted by your experimental design. If you cannot control for them, this will likely mean that you need to change your hypothesis, or change your experimental design. If you are unsure, then I would encourage you to look carefully at the experiments that others have conducted in the literature before discussing it with your advisor.

# 6

## Being Aware That You Can Get It Wrong

### Type I and Type II errors

With all the will in the world, when you are testing your hypothesis using statistics, there is a chance that you will accept your alternative hypothesis when it is not valid. This is known as a 'false positive' or a Type I error (see Figure 6.1). It is also possible that you will accept your null hypothesis when you should have accepted your alternative hypothesis, also known as a Type II error or a 'false negative'. While it won't be any fun to get a Type II error, as scientists, we should be more worried about Type I errors and the way in which they occur. This is because following a positive outcome, there is more chance that the work will be published, and that others may then pursue the same line of investigation mistakenly believing that their outcome is likely to be positive. Indeed, there are then lots of ways in which researchers may inadvertently or deliberately influence their outcomes towards Type I errors. This can even become a cultural bias that then permeates the literature.

Humans have a bias towards getting positive results (Trivers, 2011). If you've put a lot of effort towards an experiment, then when you are interpreting your result you might feel motivated towards your reasoning making you more likely to accept your initial hypothesis. This is called 'motivated reasoning', and is a rational explanation why so many scientists get caught up in Type I errors. This is also known as a confirmation bias. Another manifestation of this is publication bias which also tends to be biased towards positive results, with as many as 84% of studies supporting their initial hypothesis (Fanelli, 2010). Thus scientists, being human, are more vulnerable to making Type I errors than Type II errors when evaluating their hypotheses. Although there are deliberate or inadvertent ways of making of Type I errors, here it is important to understand that simply by chance you can make a Type I error: accepting your alternative hypothesis even when it is not correct. To minimise this possibility to a reasonable level, most biologists set alpha ($\alpha$) to 0.05 as an acceptable risk that they will encounter a false positive (Type I error).

In the following figures (Figures 6.2, 6.3 and 6.4), you see the outcome of 1000 hypotheses. You could think of these as outcomes of 1000 attempts at testing similar hypotheses in a global scientific effort by individual scientists. The power of the analysis is set at 40% and $\alpha$ is set at 0.05. It is important to realise that $\alpha$ applies to all the tests that are made (whether or not the

DOI: 10.1201/9781003212560-8

|                                    | Hypothesis is untrue | Hypothesis is true |
|------------------------------------|----------------------|--------------------|
| Hypothesis found to be correct     | Type I error         |                    |
| Hypothesis found to be incorrect   |                      | Type II error      |

**FIGURE 6.1: Type I and Type II errors.** In this figure, the columns refer to the truth regarding a hypothesis, even though the truth is unknown to the researcher. The rows are what the researcher finds when testing their hypothesis. The blue squares are what we are hoping to achieve when we set and test our hypothesis. The grey squares may happen if the hypothesis we set is indeed false. The other two possibilities are the false positive Type I error (red), and the false negative Type II error (black). In the figure, it seems that the chances of getting one of the four outcomes is equal, but in fact, this is far from reality. There are several factors that can change the size of each potential outcome to testing a hypothesis.

hypothesis is actually correct). This means that when $\alpha$ is set at 0.05, 5% of the time scientists will find that their hypothesis is correct whether or not this is actually true is, as we shall see, an interaction of several variables.

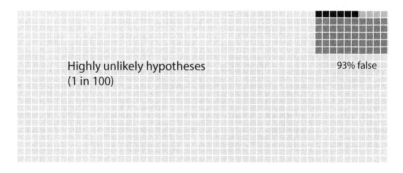

**FIGURE 6.2: The likelihood that highly unlikely hypotheses you test are accepted.** This figure has been redrawn after figure 1 in Forstmeier et al. (2017). This is a graphical representation of an argument first made by Ioannidis (2005).

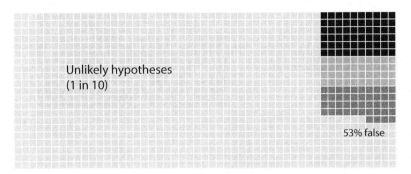

Unlikely hypotheses
(1 in 10)

53% false

**FIGURE 6.3: The likelihood that unlikely hypotheses you test are accepted.** In this figure, we see a scenario of unlikely hypotheses that are found to be correct approximately one in ten times. Now we see that the possibility of committing a Type I error is roughly equivalent to a Type II error and to the probability of finding that the hypothesis is truly correct. Thus, if your result comes out positive, you are unlikely to know why.

## 6.1 Changing the likelihood of a hypothesis

The difference between Figure 6.2 and Figures 6.3 and 6.4 is the likelihood that the hypotheses are correct changes. In the first one (Figure 6.2), we see highly unlikely hypotheses that will only be correct one in a hundred times. The blue squares denote the proportion of times in which the hypotheses are tested and found to be true. The black squares denote false negative findings, i.e. a Type II error. The red squares denote false positive findings, i.e. a Type I error. Because the hypothesis is highly unlikely to be correct the majority of squares are light grey denoting that it was correctly found to be untrue. Because $\alpha$ is set at 0.05, we can expect that 5% of the 990 incorrect hypotheses will give us a false positive (49 or 50 times). Although it might seem unlikely that anyone would test such highly unlikely hypotheses, there are increasing numbers of governments around the world that create incentives for researchers to investigate what they term 'blue skies' research, which might be better termed high-risk research or investigations into highly unlikely hypotheses. The real problem with such hypotheses is that you are more likely to get a Type I error than actually find that your hypothesis is truly correct.

Now imagine some groundbreaking research (probably published in a high impact factor journal) that finds a highly unlikely (but presumably desirable) hypothesis is correct (our first scenario of 1 in 100). Researchers from other labs try to repeat the experiment, and some of them also find that they get

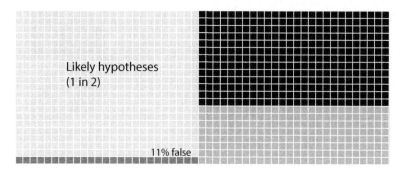

**FIGURE 6.4: The likelihood that likely hypotheses you test are accepted.** In this figure, we see a scenario of likely hypotheses that are found to be correct approximately every other time.

positive results and get these published. None of the negative results get published, or the investigators don't bother to submit them and move onto other areas of science. The positive results are written into textbooks and future generations of scientists test these types of hypotheses expecting them to be correct (our last scenario of 1 in 2). When the negative result is produced, a reasonable assumption is that a Type II error has been found, and that these scientists need to increase the power of their tests, inevitably increasing the cost of the experiment. But there are plenty of other ways that researchers could end up finding that their negative result is, in fact, positive, especially if this is what they are expecting. Simmons et al. (2011) refer to this as the 'researcher degrees of freedom' in which researchers inadvertently or deliberately end up getting positive results because of the way in which they treat either their hypothesis or analyse their data. It is easy to see how a false scientific culture can be erected in these mistaken beliefs, and it is important to be aware of this possibility.

## 6.2   Increasing the statistical power

In all of the above figures, the power of the analysis is set at 40%. The statistical power of any analysis depends on the sample size (or number of replicates) you're able to use. Some research has suggested that in most ecological and evolutionary studies this is actually more like 20% (see references in Forstmeier et al., 2017). What is important to notice in the next figure (Figure 6.5) is that when we change the power of the analysis (in this case from 40% to 20%) we influence the proportion of Type II errors over finding that the hypothesis is correct. While the overall numbers of Type I errors do not change, if your analysis tells you to accept your alternative hypothesis, there is now a 1 in 5 chance (20%) that it will be a false positive (Type I error).

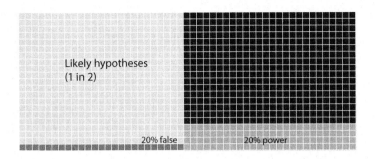

**FIGURE 6.5: The likelihood that likely hypotheses you test are accepted when you increase the power.** Here we see the scenario in which the hypotheses are quite likely to be correct one in two times. Now we can see that the possibility of creating a Type II error is highest. Next, the blue squares show us the chances that we find the hypothesis is truly correct. Lastly, there's only 11% chance of a Type I error.

What you should see when you look at these figures is that there is quite a high chance that we don't, in fact, correctly assign a true positive hypothesis. There's actually much more chance that we will commit a Type II error. Worse, the more unlikely a hypothesis is, we are increasingly likely to commit the dreaded Type I error.

From the outset, you should try to make sure that the hypotheses you are testing in your PhD are very likely to find positive results. In reality, this means that they are then iterations of hypotheses that are built on previous work. When you choose highly unlikely hypotheses, you need to be aware that this dramatically increases your chances of making a Type I error. The best way to overcome this is to look for multiple lines of evidence in your research. Once you have the most likely hypothesis that you can produce, you need to crank up your sampling so that you increase the power of your analysis, avoiding Type II errors.

In biological sciences, we have more options than many scientific disciplines to investigate potentially false cultural beliefs by using different model species, and looking at problems through different mechanisms. Presenting complementary results from multiple lines of evidence within your thesis is an extremely powerful way to build evidence for your hypothesis. However, this requires the creativity of experimental design as well as an open mind. It may mean that you have to draw upon the most fundamental of scientific abilities: to be comfortable with being wrong. It may also require dogged determination in terms of publishing negative results against the currently prevailing culture.

If your appetite is piqued by this subject, be sure to read the references cited, and read around this subject.

# 7

# What Happens If You Don't Have a Hypothesis?

So far, we have concentrated on thesis chapters or manuscripts that have hypotheses to test. I have emphasised that there is a need to focus all of the sections of your writing to helping frame and understand the hypothesis (in the Introduction), explain your approach in collecting and testing the variables (in the Materials and Methods), provide the results of your tests (in the Results), and respond to the hypothesis (in the Discussion) together with placing it in context. But what, I hear you cry if you don't have a hypothesis to test?

There are lots of reasons why manuscripts don't have hypotheses, and the chances are that in a PhD thesis (for example) you will have at least one chapter where there isn't a hypothesis tested. So, given all the emphasis above, what can be done?

## 7.1 Central problem

If you have a particular problem that your research is trying to solve, this may result in a new methodology or perspective rather than testing a hypothesis. This work is still suitable for a journal manuscript (and there are specific journals dedicated to new methodologies). Instead of introducing your hypothesis, you can simply introduce the problem that your methodology applies to. A new methodology will probably best fill an existing gap, and so your introduction will likely point out what this gap is, provide evidence for why existing methodologies don't fill the gap, and outline the variables in your novel approach.

Similarly, if you have a question but not a hypothesis, that your research tries to answer, you can take exactly the same approach to introduce your question.

Hence, no matter what your central problem is, you still need to identify this and put it at the heart of your manuscript so that all of your sections address this point. Of course, there may be more than one, and if so, you will need

DOI: 10.1201/9781003212560-9

to clearly articulate each one and explain how they are related to each other. If you are unsure about how to move forward, the best thing to do is to look to the literature for other examples of what you're trying to do. Someone has very likely done something similar before. Don't forget to chat about it with your advisor, as they may have ideas about where to look.

## 7.2   Does it matter that you don't have a hypothesis?

Having a hypothesis makes your work stronger than if you only have a question or prediction. In each case, while your results might help answer your question or confirm your prediction, they won't test the mechanism (so that you can't explain why you have the results that you do), and they may not be falsifiable/refutable. Of course, Popper would argue that if you don't have a null hypothesis, then your methodology isn't scientific.

To me, it is important for you to know exactly what your study is before you start writing about it, or even before you do the experiments. It is essential for you to frame the context of your study. The central reason why you are writing the study will lie at the heart of the manuscript. In the formulaic approach, it will be verbalised at the end of the introduction. All parts of the methodology will explain how you approached this central aim. So if you can't pin down what this is, then make sure you discuss it with your advisor.

## 7.3   Avoid HARKing

Hypothesising After Results are Known (HARKing) is the practice of creating your hypothesis once the results are in and you've done some preliminary data analyses to see what is significant (Forstmeier et al., 2017). HARKing has become prevalent in science because of the confirmation bias by journals (only accepting papers that can statistically accept the alternative hypothesis), especially those with higher impact factor. Increasing chances of confirmation bias, or a Type I error, is not desirable as you are more likely to accept the alternative hypothesis when it is not correct. Alternatively, you should seriously consider preregistering your aims to remain transparent.

# 8

## What's the Big Idea?

In the previous chapters, I've talked about the importance of having a hypothesis, and building that hypothesis in a logical framework within the introduction. The introduction serves to inform the reader about why this particular hypothesis was chosen, introducing both the dependent and independent (response and determinate) variables, as well as the presumed mechanism by which the hypothesis can be falsified (or upheld; Figure 8.1).

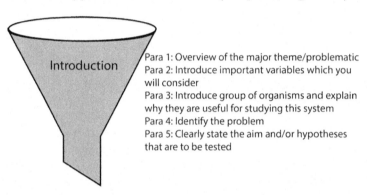

**FIGURE 8.1: A framework for writing your introduction starts with the big idea and ends with the hypothesis.** The funnel introduction is so called because it resembles a funnel in which you channel the reader from their own interest in a wider subject area into the relevance of your particular study.

## 8.1 So where would we find these big ideas?

There are quite a few papers that synthesise hypotheses in various areas of biology. Here I provide a few to get you started. Ask your advisor about relevant reviews of hypotheses in your area of biological sciences. For example, here are some theories and hypotheses in ecology (Vellend 2010; Travassos-Britoo et al., 2021; Catford et al., 2009), while these wide ranging papers about so-called 'laws' in biology and ecology (Dhar and Guiliani, 2010; Lawton 1999). Each of these papers will give you a list of big ideas, together with the citations for seminal papers that have built them. You will note that many of these theories are very old with some dating back to Darwin. You can think of the way that such ideas are structured as a hierarchy of hypotheses (Figure 8.2). Building such a hierarchy of ideas in your own specialist subject area is good reference material to prompt critical reading and reflection. You may want to take this structure forward when thinking how to introduce your PhD chapters.

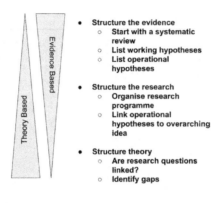

**FIGURE 8.2: A hierarchy of hypotheses can be used to help determine your big idea.** A useful structure for thinking about how hypotheses are structured was presented by Heger and Jeschke (2018) in what they termed the *Hierarchy of Hypotheses.*

Of course, there are many ways to approach and test these theories, but if you don't know about them, your work may actually make a considerable contribution to upholding or refuting them but go totally unrecognised. When the significance of your work isn't realised, it's unlikely that it'll be widely read and cited.

Let's face it, if all the effort of the work that we put into papers is just going to get buried, then is it really worth it? The work that we do is also expensive, so making it as relevant as we can to as wide an audience as possible is something that we should be concerned about.

So, I encourage you to **stand on the shoulders of giants** (Figure 12.1) by using big ideas in your introduction. Make sure that the data that you collect can actually be used to respond to some of these big ideas. Then make sure that you cite them, giving them the importance that they deserve (yes, even as keywords) so that others can find your work, and you might even find that one day, your work has shoulders that are broad enough for others to stand on!

**The take-home message**

Reading the literature can really expand your mind and broaden your horizons. When undertaking a literature review, take the time to think about not only what has been tested but what could have been. Make a list of theories and hypotheses in your own field and try to rank them in a hierarchy.

# 9

# Writing a Paragraph

Writing paragraphs has a skill all of its own, and the aim of this chapter is to go over the basics of how to put a paragraph together.

**This paragraph starts with a topic sentence which contains the principle idea that citations are needed for examples**. The following sentences put this idea into the context of your study. This should be based on examples from existing literature with the corresponding in-text citations provided (Smith et al. 2017; Jones 2018). You can use examples in your paragraph to make your point. For example, Smith and Jones (2017) carried out a study that showed that examples without citations have no credibility for readers. And this was backed up by another study by Jones and Smith (2018), which provided further evidence for the need to provide examples. Once you've provided your evidence, and introduced any covariates that you need to, like the use of literature for examples, then you are ready to sum up and link. *After reading this paragraph, the reader will know why citations are needed for examples, and they'll be ready to read more about the topic of writing a paragraph in more detail.*

In the paragraph above, the topic sentence is bold, the supporting sentences are in normal font, and the clincher is in italics.

## 9.1 Topic sentence

A topic sentence allows the reader to understand quickly the idea/topic you are putting forward in the paragraph. It must be in the context in which you are going to develop the same topic. There's no point in just mentioning a topic in passing or using it in a different way than you will later.

Make your topic sentence relatively simple. Don't be tempted to add multiple clauses. If the topic sentence is too complex, you'll lose your reader right at the beginning of the paragraph.

## 9.2    Supporting sentences

Supporting sentences convey all the relevant information to the reader. They are going to be statements that are well cited, showing readers where the original ideas came from. Be sure to keep these sentences on topic, and regularly refer back to your outline to make sure that you keep to the original objective of the paragraph. These are the meat of the paragraph and it's really important to get them right.

If you are writing about differences, then state which way the difference is. For example, if tulips are larger than irises, then say this. Telling your audience that something is different than something else only ends up leaving them guessing about the directionality. There's no point in drip dripping this information through. Set it down in as little space as possible so that your reader doesn't get bored.

Sentences within the meat of the paragraph interact. These interactions usually dwell around pertinent variables (such as those that you are going to deal with in your paper). By using the same or similar words within the paragraph, you are able to demonstrate to the reader how those different points interact.

While repeating the names of the variables, or their abbreviations, can be helpful, repeating descriptive words becomes quite tedious to the reader. Repeatedly reading the same ideas repeated over and over quickly bores the reader. Bored readers are less likely to take in what you write, even if you are repeating the same information (yes, I'm repeating myself - boring isn't it). It will also give the reader the impression that your vocabulary is very limited. These days you can do a quick right-click on a highlighted word to get a drop down list of synonyms (Figure 9.1). This can allow you to go back and replace your repeated word with: recurrent, frequent, recurring, repetitive, constant or continual. That's more than enough to spice up the paragraph. However, if you're not sure whether a word is correctly replacing another, ask a friend to read it.

Your supporting sentences are best built in the formula of an argument.

## 9.3    The magic of threes

We can learn a lot about writing convincingly in science by borrowing from storytelling, and one of the greatest lessons in storytelling involves the magic of threes (Angler, 2020).

**FIGURE 9.1: How to quickly get a synonym to make your paragraph more diverse.** An example of the synonym function in Microsoft Word (other word processors are available) – right-click on the highlighted word.

In putting together your paragraph, you can use the magic of threes to break your dialogue into three parts: thesis, antithesis and synthesis (a, b and a+b). Think of it this way: there's a standard explanation of how a phenomenon that you want to investigate interacts with the variable of your choice (thesis). Then there's some contradictory evidence that suggests that the standard explanation is not necessarily correct (antithesis). Lastly, there is your approach to investigate exactly how the variable and phenomenon interact which should explain (or at least advance our understanding) of how the variable and the phenomenon are related (synthesis). It may be that this construct should be fleshed out into three paragraphs instead of three sentences within a paragraph.

You won't be able to construct every paragraph with the magic of threes, but once you have learned this rule, you'll be able to use it to help emphasise important points in your manuscript that you want the reader to remember as they move forward. What if your list is two instead of three? Well, then you'll lose the magic, but maybe you can think it through it again? If you have two, then you normally have three by means of the combination of the two (a, b and a+b).

What follows is more advice about telling a good story: using an example and avoiding being boring: i.e. replacing those long lists with a maximum of three items.

## 9.4   Use an example

Examples are a very powerful way of conveying ideas in a short amount of space. Don't replace your paragraph with an example, but do use an example if it shows the reader just what you want. You should be able to do this in a sentence (or two), but if you're tempted to go on, it's probably not a good example. Because the first paragraph sets out ideas, it's unlikely that an example there will be a good idea there.

## 9.5   Avoid lists

I'm not a big fan of paragraphs which are simply a long list with little or no thought offered. The worst ones are where there are so many citations along the way that it's really hard to pick out what is sentence and what is citation. I understand that it's important to show precedent and that there is merit in showing how widespread an idea is over taxa or in different disciplines. You never find these lists in journals where words and/or citations are limited, which suggests that you can dispense with them.

Instead, you can avoid long lists by pulling them into themes. Once you've managed to get three themes, you'll find that you can work them into your paper time and again. For example, each theme might well be worth a separate paragraph or section in the discussion.

## 9.6   The last sentence of the paragraph: The clincher

Once you've conveyed all of the information that you planned to impart in your outline, it's time for the last sentence. This should conclude the evidence that you've provided on your topic. Try not to make it lame. For example: "This shows that little work has been done." One of my pet hates are authors who insist on telling us that "little work has been done", as if this was a justification for doing anything. Instead, make it a real clincher about why the topic is important, or how and why you will tackle it. Instead (or as well if you can), you may want this sentence to link on to another paragraph (topic), especially if flow is important at that part of your outline (see below). If you have just constructed a paragraph with thesis, antithesis and synthesis, then your clincher might be simply to underline that your approach in the paper is to investigate the synthesis. Either way, make sure that your last sentence is on topic, and one that sticks in the readers' minds.

## 9.7   Above all – read it!

Your paragraph is not finished until you've read it. That reading is an essential part of writing cannot be emphasised enough. Does your paragraph say what you planned in your outline? If your paragraph and any other text doesn't make sense to you, it sure won't impress anyone else. If you can't bear to read it through immediately, then do it after you've written two or three paragraphs. I suggest that you don't wait until you've finished the manuscript. Keep checking with your outline as you go along.

## 9.8   How does the paragraph fit into the flow?

So now we've gone over the formula, it's time to take a step back and look again at the paragraph in the context of your outline. Remember that the paragraph represents a single subject but that it is still just part of the manuscript as a whole and you need that to flow from beginning to end. This means that it's not enough to write each paragraph in isolation but to think of the way in which they link together as a whole, especially when you are trying to make a logical argument.

You should be using the outline to provide the logic in the flow of your argument, fleshed out with the relevant references. If you already know this well, then writing the paragraph will just be a question of styling the outline. Remember that there is a purpose to your paragraph. You have an argument to make the reader understand. Don't be tempted to include information that is distracting or moving away from the flow of that argument.

I am very fond of ending a paragraph on a linking sentence. Essentially, this shows how two ideas are connected in the last sentence. This really helps with getting the flow of an introduction or discussion, but linking sentences are not always the best way to end a paragraph. Sometimes there's no option but to change the subject completely, and then you should go for the clincher idea (see above). For example, you may want to end the paragraph by seeding a new twist on the paragraph's idea.

---

## 9.9   Seeding ideas

The introduction sets out the established literature in order to put your study in context, but your discussion provides you with an opportunity to present new ideas, or to turn and twist existing ideas in a new way. Once you've got a good idea of what these are, I like to seed the introduction with hints as to what these might be. Sowing seeds early in a manuscript will provide the reader with hints as to where you are going. Writing these seeds as questions is a really good way of sowing them into an introduction. You can then go on to answer them (if only partly) in the discussion. Beware though, there's no point in asking a major question in the introduction to which your data has no relevance!

---

## 9.10   Breaking the rules

Just as in the other instructions on formulaic paper writing, when writing a paragraph, you shouldn't feel totally constrained so that you can't break the rules. Breaking the rules can set you free, and much of what you read that really stands out will do this. However, it's much easier to break the rules and get it wrong than break them and get it right. The idea of this chapter was to help you get started, not to communicate with those who are already writing great stuff. So if you're already great, don't break it by changing to a formula!

---

## 9.11   Some common paragraph mistakes

- Don't start your paragraph by linking back to the previous point. Start with your new topic sentence. If you aren't ready to change the topic, you aren't ready for a new paragraph.
- Don't start the paragraph with an unsubstantiated or over-inflated claim (keep it calm and within the bounds of what you are citing). Similarly, don't

start with a trivial caveat. And don't start with a list or a long string of citations.

- Don't let your paragraph get too long. It is said that 250 words are the threshold. If you haven't covered what you wanted to in your outline within 250 words, then you need to rethink your outline for that paragraph! Definitely don't split a paragraph that's too long into two!
- If you started without an outline, it's never too late to go back and write an outline. Simply write down what you are trying to say (as a set of bullet points), and why (see below for an example).
- Your paragraph should have more than one citation in it. If you find yourself citing the same paper several times in the same paragraph, then there's something wrong with your outline, and the way that you are reading and citing papers (see below). Remember that what you write is about making your own argument from the literature, and not copying others' arguments.

The English grammar rules apply when writing science. I can't go into these in detail here as they are legion and complex. However, here are a few mistakes that are often made by those writing in science:

- You should not start a sentence with an abbreviated genus. Instead, write the genus name in full.
- You should not start a sentence with a numeral. If you have to use the number as the first word, write it out. Otherwise, rearrange your sentence.
- Make sure that acronyms are spelt out on first use (this may be needed for each section).

# 10

## Construct a Logical Argument in Your Writing

Writing is not straightforward. Your objective is to communicate with a reader, someone who you've likely never met. You need to communicate highly complex information. But more than that you need them to see things as you do. You need to provide them with your reasoning and your argument, and have it make sense to them; preferably to the point where they agree with you.

To communicate, you need to start from common ground. The beginning of your introduction starts with the most general concepts in the context in which you are writing. The context depends on your audience, and this, in turn, relates to the particular journal that you are writing for. Even if you are writing a thesis, you should pick out a target journal for each of the chapters. Once you've established the common ground, you need to carry the reader towards the hypothesis or question that you propose. The easiest way to do this is to make use of a logical argument:

A series of statements that introduce a starting premise, provide evidence for and against that premise (perhaps adding in an example that makes your point), point out what missing information would allow reaching a better understanding of said premise, and logically conclude that what you are doing is going to help fill this gap.

This logical argument style is most prominent in the early introduction of your manuscript, although the entire introduction could be seen as one long logical argument, with a few smaller more precise arguments being thrown in along the way. You might also use a logical argument in your discussion to explain how you deduce certain inferences from your results, or provide a logical extension for a future study.

The following science argumentation model (Table 10.1) is modified from Cope et al. (2013). Try using this model to guide your writing of a paragraph of your introduction (or discussion) where you attempt a logical argument. For a more in-depth treatment of using arguments in science, see Toulmin (2003).

If your logical argument doesn't arrive immediately, don't force it. Give it some thinking time to let it settle in. I find that logical arguments develop best when I'm not sitting at my desk writing. They come when I'm reading other papers (especially those with well-written arguments), taking a deep, hot bath, or when I'm out running in the mountains. If you are more social,

DOI: 10.1201/9781003212560-12

talk through your logical argument with others, perhaps over a cup of coffee. It really helps to keep the ideas turning in your head. Don't cut yourself off from work thoughts when you are away from work.

**TABLE 10.1: A framework for a scientific argument.** Modified from Cope et al. (2013), this model argument can be followed to provide as a way of getting started.

| Structure | Explanation |
| --- | --- |
| A position statement / question / hypothesis / theory / problematic | This could be the big idea in your manuscript, or one of several competing concepts that you are introducing. The context in which you are writing might mean that this idea needs no introduction (e.g. the theory of evolution in the journal Evolution), but you need to be confident that your audience will understand what you are proposing. Remember to cite the person who came up with the idea. |
| Claim 1 | One potential explanation or interpretation of the original idea. Evidence: Literature that agrees with this interpretation (could include an example) Reasoning: Your justification that links the evidence to the claim or interpretation. |
| Claim 2 | Another potential explanation of the previous interpretation of the original idea. |
| Counterclaims | Other possible interpretations or counterclaims. |
| Conclusions | Evaluation: your judgement on weighing up the evidence for the idea. |

There are other shorter forms that might suit you better. Consider another classic form of the scientific argument: "compare and contrast" which allows you as a writer to quickly familiarise your reader with some key examples. Another sentence structure to consider is explaining cause and effect. For practical wording examples of both of these classic arguments, see Part II.

## 10.1   Arguments in Latin

From time to time, you may come across comments in Latin about arguments that you've made in your work. Why do people use Latin and not simply say it in English? What do these phrases actually mean? While you may think that these people are being pompous or trying to demonstrate some scholarly superiority by quoting Latin at you (and you might be right), there is some important reasoning to consider both in the meanings of these *argumentum* (as they are known in Latin), and the fact that they are as old as Roman scholars, and often much older. This then is the point, that some arguments are faulty, and that this has been known for thousands of years.

Here I have provided some of these arguments, and the reasons why they may appear in your thesis in biological sciences.

*Ad hominem* – "To the man" means that the argument attacks the person and not the work.

*Non sequitur* – "It does not follow": suggests that one statement in your argument is not linked to the next. This is also known as a logical fallacy, a formal fallacy or a deductive fallacy.

*Post hoc, ergo propter hoc* – "it happened after, so it was caused by": mistaking cause and effect simply because one thing happened after another. The two events may have been completely unrelated.

*Reductio ad absurdum* – "reduction to absurdity": attempts to establish a claim by comparing it to an absurd opposite.

*Novacula Occami* – "Occam's razor": The idea that the simplest hypothesis, or argument is most likely to be correct. Another way to think of this is if you have two models that you test are equally likely (e.g. within 2 $\delta$AICs), the one with the least number of variables should be considered superior. This is also known as the principle or law of parsimony: *lex parsimoniae*.

---

## 10.2 Other arguments in English

Likewise, there are some arguments in English that you will hear time and again. I take the opportunity to also list these here:

**Correlation does not equal causation** – That two events or variables are related when in fact, they are co-incidental.

**Absence of evidence is not evidence of absence** – The idea that positive evidence of the existence of a relationship is superior to a lack of the same evidence. This is also known as a Type II error (false negative), where the lack of finding a significant relationship does not lead to a conclusion that the relationship does not exist. Later we discuss how a lack of finding of a significant result leads to confirmation bias in science.

**Observational selection** – "The root of all superstition is that men observe when a thing hits, but not when it misses". Francis Bacon (1620). This refers to the cherry picking of studies that support your hypothesis while omitting those that do not support it. Later we see how this can be abused when writing the introduction to your chapter.

**Straw man** – Erecting a false or simplified statement in order to make it easier to attack.

**False dichotomy** – Considering that two extremes of a continuum are the only possible alternatives.

We have seen elsewhere how science is built on the works of those that have worked before us. When we construct a logical argument in science, we do so using this scholarly accumulation of knowledge as presented in the literature: citations. That is to say that your claims must be backed up by the literature. To do this, you will need to read that literature and make sure that it can back up your claims. Beware of making a baseless claim.

It may be that when researching your question, you come across the same argument using the same literature rehashed time and again in different papers (it happens). Does this give you a green light to do it again? I'd like to think that you already know that there are likely lots of other untapped and better examples out there, and it'd be well worth your while constructing your argument yourself.

Beware of copying an argument wholesale. You might well end up getting muddled, or worse perpetuating an error. Better go back and make sure that you understand the original premise and the works that promote or oppose this. Having told you to beware, I'm now going to encourage you to read, because reading is one of the best ways in which you can learn about writing a logical argument.

Reading critically will make you aware of when you come across an argument. Ask yourself:

- Did you understand it?
- Was it written in a conventional style (as in Table 10.1)
    - If not, how was the style broken, and did this improve or detract from the understanding?

In this writing book, I concentrate on providing formulaic approaches to writing because these are by far the easiest ways for inexperienced writers. However, I encourage you to learn and experiment with writing styles as you become more experienced. Critical reading is one of the best ways to learn about alternative writing styles: reading is probably the best way of improving your writing. Alternatively, and especially if the above hasn't clicked with you, you can read more about writing a scientific argument (see for example here[1]). And there's plenty more out there.

---

[1] https://msu.edu/course/iah/231b/gifford/sciaargs.htm

# 11

## Storytelling in Science?

I started thinking about this topic some years back as I often need to write popular articles that make some of the science that we do more generally accessible. But then I came across an article by Nick Enfield (2018) that made me think again.

'No storytelling' is a comment that I sometimes make when reading drafts of manuscripts, chapters, and even when editing for journals. What do I mean by this? Stories are deterministic. That is to say that the storyteller has an end in mind when they start telling the story (i.e. teleological; see Figure 11.1), and the telling is a way to get to their goal. A story that's 'pointless' will frustrate the audience and won't engender them to listen to that storyteller again. In a good story, reaching that goal will often result in lots of twists and turns with the goal shrouded in mystery until it is revealed. In a teaching story (like a parable), the goal may be overt, such that the audience relates to the narrative and buys into the same conclusion.

If we did science like we tell stories, we would decide on the way the system works before we studied it, and then design the experiment in order to reach our desired goal. You should have recognised by now that this is not the way we do science. This is clearly an undesirable way to go about doing science because we should never prejudice the result that we'll get from a study before we do it. Hence, writing stories and writing science are different processes.

## 11.1   The danger of a teleological argument

Teleological arguments are those which have their endpoint as the aim. While it is generally acknowledged that this is a good way to tell a story, it can be a trap when it comes to conducting scientific inquiry. Here, we need to approach science in a very different way to storytelling. When I was doing my PhD I was very frustrated as I had the impression that my advisor knew what result he wanted and designed the study to show it. What probably happened is that he was displaying 'motivated reasoning', but the result is known as

DOI: 10.1201/9781003212560-13

**FIGURE 11.1: It's story time.** Are you sitting comfortably? Then I'll begin.... but it won't be science.

'confirmation bias'. When an experiment failed to meet the expected result, he declared that it had failed. As Frances Bacon so elegantly put it:

> "Once a man's understanding has settled on something (either because it is an accepted belief or because it pleases him), it draws everything else also to support and agree with it. And if it encounters a larger number of more powerful countervailing examples, it either fails to notice them, or disregards them, or makes fine distinctions to dismiss and reject them, and all this with much dangerous prejudice, to preserve the authority of its first conceptions."
>
> Francis Bacon (1620)

In our work, we read other studies and observations to formulate a question that we frame as a hypothesis. We then devise an experiment that will test this hypothesis in the most objective way so that we can fairly accept or reject our null hypothesis.

---

## 11.2    Science is the very opposite of storytelling. Or is it?

I often tend to think of the answer to the question as the goal of the study. That I don't know what the answer is, doesn't spoil the story for me. The important thing is asking the right question. An unexpected answer might send us back to thinking more about the system that we are studying and result in a greater revelation.

A good example of this is the study published by Becker et al. (2018). Francois found a (spookily) strong relationship between probability that Rose's dwarf toadlets survive and local rainfall in the breeding season. More rainfall equals greater survival is what one would automatically think (because frogs like rain), but that wasn't the result obtained. Francois found that survival increased with less rainfall. It wasn't until we made the connection between the fact that increased rainfall during the breeding season meant that the toadlets spent more time in puddles, that we understood why their survival was decreasing. As the toadlets aren't feeding during this time they lose mass, and also expose themselves to increased predation. In a dry year, the toadlets will head back to their subterranean refuges much earlier and continue to pursue 'safer' feeding and hiding habits. While we might intuitively feel that a 'safe toadlet' is a better life strategy, reduced rainfall means reduced reproduction, and so results in a 'failure' for toads that don't manage to pass on their genes that year. The result is an extremely variable life history with the weather, something that was previously unknown. To me, that's a great story!

But the scientific paper that was written about this 'story' (Becker et al., 2018) doesn't have a narrative style, and doesn't fit the description of a narrative that is given above. Instead, it follows the conventional formula: Introduction, Materials and Methods, Results and Discussion. This style does not treat the experiment as a complication on the way to the story's goal. The structure introduces the rationale behind conducting the experiment, then objectively explains the findings, and lastly discusses their meaning in relation to what is already known. From this study, I can spin the story (above) because I know what the endpoint is. But this story is just one interpretation of the correlation that was found. Correlation is not causation, and more studies would be needed to test the hypothesis concerning the underlying mechanism suggested for the relationship observed. The key differences in the storytelling style from the scientific formula are the absence of a known goal at the start of the process (see a more detailed discussion of this by Katz, 2013).

There is a role for stories when communicating science as this increases interest, facilitates understanding, and enhances memory. This is particularly true when communicating science to the public and making it more accessible, but it also applies to interactions between scientists, for example, at conferences. The presentations that tell a story and entertain are those we tend to remember. Not easy, but if we do want to communicate well with each other, then we need to learn the art of storytelling without compromising our scientific objectivity. In addition, there are lots of lessons from storytelling that we can use within scientific papers, and to which I subscribe wholeheartedly. For example, don't be boring, don't bamboozle, write an active narrative (not in the passive tense), use the magic of threes, and many more.

I've taken the deterministic interpretation of storytelling (a story with a known ending), and I've used this to argue why it is not a good tool for science. But it is a great tool for communication. And storytelling can be used to improve the way that we write science, and especially the way in which we communicate science.

There are other fundamentals in stories that are particularly useful and should not be ignored. One notable feature of stories is that they make facts easier to understand and remember. Indeed, there is even the idea that this is why storytelling evolved in human societies. For example, most religions are based on stories that are geared to understanding of societal morals and rules. Martin Angler's (2020) excellent book about how to turn science into stories should be consulted to give you some deeper insight into why stories are more powerful than science. Or try the exercise described by Torres and Pruim (2019).

## 11.3  Can storytelling help memory of science?

Some argue that storytelling can and should be an integral part of scientific writing. In her blog post[1], Anna Clemens argues that stories should be used within the scientific context. I'll let you read this for yourselves, and ask you to make up your own minds. If you like this way of thinking, you can get deeper into the depths of story thinking with Angler (2020).

---

[1] https://www.annaclemens.com/blog/elements-of-a-story

# 12

## Why Do You Need to Cite?

Citations are a means of recognising that your work is built on the work of others. They are part of the fundamental tenets of science; that we don't start from scratch each time we conduct a study.

# Google Scholar

```
[                                    ] [🔍]
```
● Articles ○ Case law

**Stand on the shoulders of giants**

**FIGURE 12.1: Google Scholar home page.** Have you ever wondered what the blurb on the front of Google Scholar means? Who is standing on whose shoulders? Google and the Google logo are registered trademarks of Google LLC, used with permission.

---

Isaac Newton (in a letter to Robert Hooke in 1675) wrote:
**"If I have seen further it is by standing on the shoulders of Giants."**

---

Although the phrase is often attributed to Isaac Newton, it turns out that it was well accepted in Newton's time and has been traced back to the 12th century.

DOI: 10.1201/9781003212560-14

Essentially, standing on the shoulders of giants (Figures 12.1) is a recognition that all research is built upon research that has gone before it, and this is the basis for citations in the text of scientific papers. Patrick Dunleavy (2017) argues that citations are required to meet seven criteria with respect to academic writing. Others have suggested that there are more criteria (see Harwood, 2009), and that these are a fundamental premise of academic work. But it was not always thus. If we dwell a moment longer, we can look back at the World's longest-running scientific journal: *Philosophical Transactions of the Royal Society.*

You can download the first of those papers of 1665 and note that there are no citations (other than to books or letters) because there were no previously published articles from which to draw (see Figure 12.2). However, even then, authors noted that ideas came from previous authors and we can regularly find acknowledgements that refer back to Aristotle (~350 BC).

## (10)
### *An Account of a very odd Monſtrous Calf.*

By the ſame Noble perſon was lately communicated to the *Royal Society* an Account of a very Odd Monſtrous Birth, produced at *Limmington* in *Hampſhire*, where a Butcher, having cauſed a Cow (which caſt her Calf the year before) to be covered, that ſhe might the ſooner be fatted, killed her when fat, and opening the Womb, which he found heavy to admiration, ſaw in it a Calf, which had begun to have hair, whoſe hinder Leggs had no Joynts, and whoſe Tongue was, *Cerberus*-like, triple, to each ſide of his Mouth one, and one in the midſt: Between the Fore-leggs and the Hinder-leggs was a great Stone, on which the Calf rid: The *Sternum*, or that part of the Breaſt, where the Ribs lye, was alſo perfeᷓt Stone; and the Stone, on which it rid, weighed twenty pounds and a half; the outſide of the Stone was of Greniſh colour, but ſome ſmall parts being broken off, it appeared a perfeᷓt Free-ſtone. The Stone, according to the Letter of Mr. *David Thomas*, who ſent this Account to Mr. *Boyle*, is with Doᷓtor *Haughteyn* of *Salisbury*, to whom he alſo referreth for further Information.

FIGURE 12.2: A page from the first volume of the Philosophical Transactions of the Royal Society. Boyle, R. (1665). An account of a very odd monstrous calf. *Phil. Trans. R. Soc.*, 1(1):10–10. https://doi.org/10.1098/rstl.1665.0007

## 12.1   Research is built on existing work and ideas

It would therefore be very unlikely that your idea has no basis in existing literature. If you can't find it, the chances are that you haven't looked in the right way. Try Google Scholar, Scopus, Web of Science and vary the search terms or try searching for articles citing something similar.

Citations demonstrate to readers where your ideas have come from. Citations can also be used to reduce what you need to write – especially with respect to methodology. If you (or others) have already provided the methodology in full then you can give a much simpler description and the citation.

Citations need to be used to back up any statements of fact that you make in your PhD chapter. Any examples need citations for where the examples came from, and when you make arguments, you add credibility to your arguments from each side by citing that these are from published works.

## 12.2   Your citations say a lot about you

When you choose to cite one paper over another, you are making a statement about what you find credible in the literature and what you don't. For example, if you choose to cite a poorly conducted study as the cornerstone for one of your arguments, then others may interpret that your arguments are built on shaky foundations. Experienced (often senior) readers who know the literature well will be able to judge the quality of your work not only by what you've written but on what you've based your ideas through your citations. As we will see later on, there's a lot to think about when deciding what to cite, and what not to cite.

## 12.3   Cite while you write (not afterwards)

Automatically, I'd suggest that you cite whenever you write. Using citations becomes habit forming, and you'll end up wanting to use citations everywhere. Popular scientific writings tend not to have citations, they can still be there but subtly different.

Within a manuscript for a scientific journal or in your thesis, you can expect that your introduction and discussion sections are going to be full of citations.

The methodology will likely also be citation rich. By using a lot of citations in your methods, you can often save your needing to write a lot of sections which would otherwise be very detailed.

Later in the book (Writing with a Formula), I explain how to start writing with an outline, then flesh out the outline with citations from your reading, and plan which citations you are going to use carefully before you start. This won't stop you from adding more later, but this ensures that your work is built on your scholarly endeavour.

I've seen many students try to write their text first and then only later fill in citations. I've had students hand in thesis chapters and manuscripts that have '[ref]' written strategically at the end of many sentences. I've even had the experience of doing this myself when I was an early career researcher (I probably still have those manuscripts somewhere because it's a sure way of never getting anything finished). I suggest that you **do not do this.** The reason is that if you write first and try to insert citations later, you are going to spend lots of time looking for the citation that fits your text. This is clearly not the right way around. Your text should be based on what you know to be present in the literature.

I was planning to write that no one would ever write an entire manuscript without references, and then put them all in at the end. Then I came across a video by Pat Schloss[1], and realised that I'm wrong (Schloss, 2018). Pat states early on (in the video) that he prefers to write first and cite later. He's clearly a very experienced researcher and knows his literature very well. Indeed, he knows it all so well that he already knows what citations fit what statements without needing to look them up. However, later on in the video, Pat admits that even he sometimes gets it wrong, and that means that he ends up having to rewrite text when he finds the publication that he was thinking of and notes that what they actually found did not fit what he wrote. Although I can identify with what Pat is saying, I would find this re-writing far too time consuming to take this approach myself. I am familiar with most of the pertinent literature in my specialist area, but I'd rather check things before I start writing than leave it until later. I find that my memory is not only wrong about what other people have found but also about what's in my own work. Therefore, I'd suggest that while you may aspire to write like Pat later on in your career, right now it'd be better for you to plan your writing with citations in mind, and cite while you write (not afterwards).

When you submit a draft of your chapter to your advisor, they may suggest additional citations that are known to them. Don't simply add them blindly, but look them up and check their relevance to the statement concerned. They may require recrafting of the statement, or a query back to your advisor.

---

[1] https://youtu.be/yk54_j3-4RM

## 12.4  Citation styles

There are a number of different styles, and this is likely to depend on the journal that you are writing for (see Pandey et al., 2020). The two most prevalent styles in biological sciences are often referred to as Harvard (name, year) and Vancouver[1] (superscript numbers that are listed in the reference section). Most universities and journals require Harvard style. The intricacies of how exactly this will get carried out will change from institution to institution and journal to journal. You will have to find out what is relevant to you.

### 12.4.1  Vancouver style

Vancouver style dates back to medical journals in the 1970s and refers to conventions to avoid fraud in medical sciences. You are probably familiar with it, even if you don't know the name. It is the convention that uses small superscript numbers to denote the place of citations in the main text, and the full reference given at the end of the document is provided in order of the citations (and their numbers) instead of alphabetical order. In biological sciences, Vancouver style is not used as widely as Harvard style.

Each style has its pros and cons. Vancouver style is more equitable and anonymous, as all cited people are represented only in the references and do not have their names plastered throughout the text. The diminutive numbers take up less space, allowing the reader and writer to concentrate more on their prose. On the other hand, if you are familiar with the literature, or you are interested in learning more about the literature, you'll find that the little numbers take much more time to cross-reference with the list at the end than does Harvard style. Reference strings in Vancouver style can either be separate numbers, or numbers listed as a series. Because citations are numbered in the order in which they are first mentioned, citation strings later in documents often end up as a list of non-sequential numbers separated by commas.

Here is an example of text (from Measey et al., 2016) with citations in Vancouver style:

Amphibian populations are currently declining across the globe[1-3] and alien amphibians are at least partially driving these declines through competition[4], hybridization[5] and introduction of novel pathogens[6-9].

The corresponding references in Vancouver format are in the section on references.

## 12.4.2   Harvard style

The standard way is to use the name and year in parentheses at the end of the statement to which the citation is relevant. Names of three or more authors are frequently reduced to 'et al.' (often written in italics: *et al.*) which is short for the Latin *et alia*, meaning 'and others'. References strings in Harvard style can go on for an entire line of text or more. Some journals have rules that mean on the first mention, the citation should have all authors (up to a certain number). This can become tedious when citations start taking up more space than text.

A repeat of the above example, except with Harvard style shows how much more space these same citations take up.

---

> Amphibian populations are currently declining across the globe (Wake and Vredenburg, 2008; Collins et al., 2009; Pimm et al., 2014) and alien amphibians are at least partially driving these declines through competition (Kupferberg, 1997), hybridization (Dufresnes et al., 2015) and introduction of novel pathogens (Berger et al., 1999; Daszak et al., 2003; La Marca et al., 2005; Martel et al., 2013).

---

You can find the corresponding references in Harvard format in the references section.

From here on (and throughout the book – because the book uses a form of Harvard style), I concentrate on Harvard style as this provides more freedom for how citations can differ.

---

## 12.5   Moving from Harvard to Vancouver

Moving within styles is always a hassle, especially as formatting the references can take a long time, but moving between these styles is especially tricky as it may require you to re-evaluate the way in which you use the citations.

Moving from Harvard to Vancouver style means that every time you would have had the authors and year, you will replace this with a superscript denoting the number for that particular citation. However, Vancouver style does not allow for much variation in how you cite. You shouldn't use (e.g. [1,2]) or (see [3] for a review). This means that when you originally wrote for a Harvard style journal but then change the manuscript to submit it to a journal that uses Vancouver style, you'll need to remove any explanations of citations. That doesn't mean removing the citations themselves, but it may mean re-writing some sentences. Simply changing your reference manager to use a different referencing style will only be the start of your work.

## 12.6   Where within a sentence should the citation come?

There are a number of different styles, and this is likely to depend on the journal that you are writing for. The standard way is to use the name and year in parentheses at the end of the statement to which the citation is relevant.

> The impact of all invasive amphibians is similar to that of invasive birds and mammals (Measey et al., 2016).

You'll see that sometimes the names are brought into the sentence and become the central agents of the text.

Sometimes, instead of '*et al.*' you can write 'and colleagues' or 'and others'. This is something to do occasionally when you are looking to diversify some text. Don't over do it though.

> Measey et al. (2016) found that the impact of invasive amphibians is comparable to that of birds and mammals.

This technique is very useful when you then want to add another sentence or two about this same study.

Measey et al. (2016) found that the impact of invasive amphibians is comparable to that of birds and mammals. They did this by constructing GISS scores for all species in all groups.

Because the authors are the subjects of the first sentence, the citation becomes implicit in the second sentence. Then you don't need to use the same citation again within the paragraph.

What about page numbers? Sometimes you'll see a citation with a colon and page number after. This really only needs to be used if you are quoting specific text on a particular page:

Measey et al. (2016, 976) proposed that using GISS scores could show that "some amphibians can have devastating impacts on the environment".

## 12.7   What about the order of the citations in a string?

A citation string is a list of two or more citations that all assert the statement given. When you provide a string of citations, you will need to decide which one comes first. Normally, this is given in the order of precedence: the oldest citation comes first, the youngest last. However, journals have their own styles and these may dictate how citation strings are ordered. For example, they could be alphabetical. If your referencing software has a style then you can relax. Otherwise, you'll need to look it up!

## 12.8   What about citations as taxonomic authorities?

Taxonomists have special rules for this, and this will be explained in another chapter (see section on Scientific names and taxonomic authorities). These are not the same as regular citations (because they don't appear in the literature cited and you don't have to have read the descriptions), and only some journals ask for them.

## 12.9 Is it possible to mis-cite?

Yes. One of the most common ways in which students mis-cite a paper is to use statements made in the introduction (or discussion) which were not the subject of the study. For example, in the introduction of their paper, Measey et al. (2016) make comments on amphibian decline.

However, it would be wrong to give a statement on amphibian decline and cite Measey et al. (2016). They did not study amphibian decline. Instead, you should read the papers that they cite (e.g. Wake and Vredenburg, 2008; Collins et al., 2009; Pimm et al., 2014), and read around those to find studies on amphibian decline that are appropriate for your context. This underlines one important aspect of choosing citations where the statement that you make relates directly to the study carried out in the citation.

Another common mistake is to forget which paper has which information. You can try to make sure that you don't do this by taking better notes or write a more accurate outline. And do the citations before or as you write, not afterwards!

## 12.10 Should I cite without reading the paper?

No. When you are citing a study, you should be sufficiently familiar with the publication as you are endorsing the study in relation with the statement that you make (but see below). If you are not convinced by the nature of the study that you are tempted to cite, then rather don't cite it and use another one. If you can't get hold of the paper, this is another reason why you might not cite it. This is a regular reason given for why Open Access journals attract more citations.

## 12.11    What should I not cite?

This does depend on the journal you are writing for. Some journals don't
permit citations to unpublished data or websites. My suggestion would be
to avoid such sources anyway (with particular exceptions – see below) unless
it is really important that you include it. Other examples of texts to avoid
are: textbooks (use the original paper instead), newspapers or magazines, blog
posts, Facebook pages (or other social media sites), predatory journals (or any
non-peer-reviewed article). These are all examples of **grey literature**.

There are, of course, exceptions. When you are writing about social media
sites or newspaper coverage, you will want to cite those sources.

I don't like citations to guide-books or general text books. This has become
very common but really smacks of laziness. Most of what's written in a guide-
book has already been documented in the scientific literature, but guide-books
generally don't provide sources for the information that they provide. Thus,
it's easy to look up and find something in a guide-book but suggests (to me)
that you haven't spent enough time or effort reading the literature. Some
guide-books are excellent and the authors have gone to a lot of trouble to in-
corporate original data and observations. But this is unusual, and most data
can be found in publications.

One exception to not citing websites relevant to our work is the IUCN Red
List. Note that all entries on this site now have DOIs, and this might be a
good guide for what is available to cite. The Digital Object Identifier (DOI)
is very useful as it means that there is a consistent record of that version.
Otherwise, you could cite any website and then the owner can go and change
the site and it no longer says what you thought it did. The DOI removes this
problem; there will always be an archived version with that particular DOI.

You should not cite papers that have been retracted.

## 12.12    Do I cite the review or the primary literature?

The primary literature consists of studies or experiments that are done in order
to test a hypothesis. Secondary literature includes reviews, syntheses or meta-
analyses. Primacy (see below) is important, but this depends on the space
you have and whether the review contains all the information you need to cite.
Sometimes reviews (especially meta-analyses) provide more information. It's
preferable to use primary literature, but sometimes reviews (or meta-analyses)

are actually more expedient to use, especially if they are not the focus of your study. You can even cite both when relevant.

## 12.13   What is primacy and why does it matter?

Primacy in the literature is the paper/s where the authors provide the original idea or concept that you are citing. It is important in that you should give credit to original ideas over those who copy or simply repeat them, or even those who review them. Sometimes primacy is less important, especially if the concept is well known and has changed substantially. Then you should cite the most recent work that shows the ideas that you are wanting to show. Other times, primacy is important – especially when few have built on ideas or concepts since they were put forward. If in doubt, place a citation to the original study and the most recent study, then flag this using a comment for your advisor to consider. The important point here is for you to give credit where it's due and not to overlook those who put in the hard work to publish original content.

## 12.14   How many citations are enough?

Some journals have a word limit or even a limit to the number of references that they allow. Others do not, and you should probably use what is recently published as a guide to what is acceptable. For the chapter of a thesis, you should err on the upper end, from 50 to 100 references. Note that citations may well be more as you may cite a paper more than once.

Obviously, everything you cite needs to be in the References (or Literature Cited) section, and you may well need to spend time deleting extra stuff. You can get around this common issue by using a citation manager like Mendeley, Zotero or EndNote. I'd suggest that you use one of these tools as they can really help with your reading too. These days they are busy turning into a kind of scientific, social network. They regularly make suggestions of what you could be reading based on what you read. This can be useful.

If you are looking for multiple examples of citations for a statement that you've made and there are many possibilities, I'd suggest that you aim to produce three. Make sure that you use a suffix (like 'e.g.') to show that you are aware that these are examples of a widely reported phenomenon. You can choose these as you like but may want to consider using what you consider to be 'the best' examples, and/or references that you are planning to use elsewhere in your paper/chapter. This can drive the total number of citations down considerably and helps to keep the citations more relevant to your work.

## 12.15   Should I cite myself?

If you are publishing relevant and appropriate papers, then there is no reason not to cite yourself. In many cases (such as with your thesis work), your own publications are likely to be more relevant to methodology and subject matter than much of the other work that is out there. However, if you've previously published on termite fungus and now you're publishing on rat toes, it's unlikely to be relevant. Choosing citations should include your comfort with being transparent. If you know when citing yourself that citing another article would be more correct in showing where an idea came from, but citing your own is more relevant to the application, then you should feel comfortable in citing both.

I would avoid citing your thesis if possible, rather put it all into papers. There are times though when citing your thesis is unavoidable. Within your thesis, I would suggest that you do provide citations to different chapters as this will help the examiners see how the chapters relate to each other.

Take a look at this article on self-citation. They claim that self-citations in the Natural Sciences run at 33% (van Raan et al., 2007), which provides you with a nice idea of what is acceptable. Remember that this would include not only your citations to your articles but all citations to your co-authors' articles too.

## 12.16   Should I cite my friends?

It may be easier to cite your friends if you already know their work well. You may have heard them talk and know that the subject is relevant. They may be encouraging you to cite them, but should you?

In these days of scientometrics, we do need to acknowledge that citations act as a kind of currency. They count towards your H-index (Hirsch, 2005) and this can reflect on your prospects as a postdoc or employee. What's also clear is that cited papers get cited more, so it could really help your friends if you cite them. Obviously, the inverse is also true, so beware of the politics of citing.

However, the most important points have already been raised. The study must be relevant and appropriate before it gets included as a citation in your work.

## 12.17 Does the impact factor of the cited article matter?

Papers in journals with high impact factors are more likely to be cited because their contents are already thought to be of interest to a wide range of people. Indeed, there is evidence that for identical statements published in many journals, those with higher Impact Factors are cited more (Perneger, 2010). Sometimes (but not always) the impact factor of the journal can be an indication of the quality of the study. But you should judge this for yourself when you critically read the paper.

I find that the first paragraph of a paper is more likely to contain citations of higher impact journals. This is, in part, as these are likely to be more cross cutting (as is often the case for the first paragraph). In the end, if you need to make a choice, choose the paper that is most relevant, irrespective of impact factor.

# 13

## Literature Databases

Databases of scientific literature have been around since the 1960s, when the number of journals and papers started increasing so much that scholars simply couldn't keep up. Quite simply, you should use one or more of these databases to look for relevant literature to read. The first database, and one that continues to be one of the largest, was the Science Citation Index (now renamed Web of Science[1]) that sought to record not only the contents of selected journals (titles, abstracts, authors and other metadata) but also the citations that each paper used. As we have seen in the last chapter, citations record the way in which people find the literature useful, and are therefore, a highly effective way of finding related and pertinent literature in your field.

Before you start using a literature database, you will need to have some key literature with which you are familiar and that relates directly to your project. If you don't have this, then check with your advisor. Sometimes, this key literature is cited in the advert for the PhD, the grant application that funded it, or if this is a topic that your advisor has already conducted some research on, then their own publications may include part of the key literature. If you are not familiar with your advisor's (including co-advisor's where appropriate) papers, then a great first step is to search for them on a database and read through titles and abstracts that are relevant to your research topic. Once you have a set of key literature (this might only be 5 to 10 papers), then it is time to go to the literature databases to do your search.

## 13.1   Searching the literature using a database

There are two typical ways in which you can search for literature about your area of interest. The first is to use keywords and the second is to use citations of key references. I suggest that you use both of these tactics when doing your initial literature search. Further, I'd suggest that it's worth your time keeping a record of what you have searched for, how and when and then download the

---

[1]https://www.webofknowledge.com/

results. This is called a systematic literature review. Using this tactic from the outset of your interactions with literature databases will give you a good idea of what proportion of the literature you are reading, and you might then be able to convert your finding into an objective literature review – making a great first introductory chapter for your thesis.

---

## 13.2   Keywords

While searching for keywords sounds easy, picking the keywords that you should use is more difficult and requires a lot of practice with the databases and your key references. Remember that keywords are often not single words but also include short phrases. Look at all of the key references and the keywords that they contain. Any keywords that they share are going to be important for you. Make a note of all relevant keywords for your searches, as ultimately, this will help you when you need to determine the keywords for your own chapters.

Most literature databases have the option of using just keywords, or keywords and words in the title and abstract. I would start with the broader search term and see how well they work. However, these searches alone may return such a large amount of literature that you can't possibly read it all. In this case, you will need to use more than a single keyword. The trick is to find the correct combination of keywords that will deliver all (or most) of your key references, while not delivering returns of thousands of papers that are not relevant to your study. Ideally, you want a search term that will deliver around one hundred papers at most, where the majority have relevance to your study area. Luckily, literature databases have some tricks for you to combine your keywords in ways that you can include or exclude certain topics.

### 13.2.1   Boolean terms or operators

There are three principle Boolean operators: AND, OR and NOT.

You should use these to combine a string of keywords. They are relatively straightforward to use:

- Use AND when you want both terms included. For example, *performance* AND *lizards*. In this example, all papers that mention performance AND lizards in the title, abstract or keywords are returned.
- Use OR when you want to include either term. For example, *performance* AND *lizards* OR *geckos*. This example should return more literature than the

previous search as it will include all papers that mention *performance AND lizards OR geckos* in the title, abstract or keywords are returned. You will note that geckos are lizards, but people who write about geckos might not mention that they are lizards in the title, abstract or keywords, so you may need to use these extra search terms in order to pick up all of the relevant literature on lizards. If your research requires you to cover all lizards, then you might need a lot of search terms in the OR string in order to pick up all of the literature that covers lizard performance.

- Use NOT when you specifically want to exclude a keyword. Following our example above, the search term *performance AND reptiles NOT snakes* should give you all performance papers on lizards, but you may want to exclude turtles and crocodilians as well if they have a lot of hits. NOT is a powerful operator and needs to be used with caution, as it could exclude some relevant hits. In our example here, an item about interactions between snakes and lizards would be excluded, as would any paper that included work on lizards and snakes.

### 13.2.2 Proximity operators and parentheses

You can also use the term NEAR to indicate that two terms need to occur within a specified number of words. For example, *performance* NEAR/5 *reptile* will capture items where these terms occur within five words of each other, but no greater. This might be useful to exclude some papers that mention performance and reptiles in the abstract but are not about reptile performance.

Once you start putting together search strings you will need to use brackets to make sure that the correct terms are grouped. For example, *performance* AND (*lizards* OR *geckos*) will get what we described above, whereas (*performance* AND *lizards*) OR *geckos* will give you all papers on lizards and performance plus all papers on geckos (not what we wanted).

Explore the advanced search options of the specific database that you are using to see the correct semantics needed for a search.

### 13.2.3 Wildcards

Most literature databases allow you to search using different wildcards. These are *, $ and ?. These can be placed at any point within a search term and help overcome differences in spelling, or families of words that you may want to capture or exclude in your search.

- Use * when you want to replace a set of letters at the end of a word. For example, _reptil*_ will return all terms including Reptilia, reptile and reptiles.
- Use ? when you want to search for any single letter at that place in a word. For example, *sterili?e* will capture words with sterilize and sterilise.
- Use $ when there is more than a single letter that might change within a word, or the omission of a letter. For example, *colo$r* will return both US and UK spelling, color and colour, respectively.

### 13.2.4   Combining searches

Lastly, there is the option of combining searches so that you can add search strings together or keep them apart. You will need to use the advanced search options in your database to do this.

You will need to carefully check your database for their precise semantics when using these Boolean operators in a search string. They are all slightly different, often using quotes around the keywords, and your search won't work as you intend unless you are using the Boolean operators in the way the database dictates.

Your specific search term may end up having a long list of terms. Remember to save this whenever you make a search, especially if you refine it during your studies. Most literature databases will provide the search string that you used together with the results. Check your search string carefully for possible spelling errors that could influence your search.

## 13.3   Moving items into your reference manager

An invaluable way in which databases have become extremely useful is the ability to move items that you have found during your search directly into your reference manager. Beware that these regularly come with errors, and that the amount of errors is related to your database source (Scopus and WoS are much better than Google Scholar). This means that after importing the references, you will need to go back to your reference manager and edit them or they will come out incorrectly when you use them. Of particular note to biologists is the need to have all species names in italics, and you'll need to do this yourself.

## 13.4   Citation searches

In addition to using keywords, you should search citations of your key references. Typically, when you start reviewing the literature in your specific area, you will have a few key references that have been given to you by your advisor. Having read these, you will already know that they cite other works of interest, and you should follow up to look at all that seem relevant. But that's not all, these key publications are likely to have also been cited themselves by newer papers. Tracking citations of key papers can help you keep abreast of what is happening in your field. Indeed, given the prolific ever increasing nature of scientific publications, it might be the only way to keep on top of what is happening. Some literature databases allow you to receive alerts if a paper is cited, and you should set this up for all of your key references, and perhaps for some of the most prominent authors in your field.

You can't search every paper for all the references that it cites and its own citations. Although this would give you a solid systematic basis for your literature search, it is likely to be too much work. Getting good search terms for your keywords, and finding key references to search for citing articles is a better strategy as it makes your literature searches manageable. If you refine and save your search terms it also makes it repeatable. This will provide you with a solid basis for moving forward with your PhD, and with appropriate alerts make sure that you keep up with the latest literature in your field.

The above section is a brief guide, and your library is likely to have in-depth guides on how to use the specific databases that they house. They may also provide training sessions, or online videos to watch. It is worth becoming familiar with exactly how the databases that you will use work, so this is time well spent at the beginning of your studies.

### 13.4.1   What else are literature databases used for?

The Science Citation Index used to be published as a single volume that emerged each year and was available in the reference section of the library. By the time that I started at university as an undergraduate in the 1980s, it was several volumes large and my first task for my tutor (one Dr Dave J Thompson) was to look up all of his citations (thankfully not too many at that time). Within a couple of years, the database was digitised and available on CD, and eventually became accessible through the internet.

Citations provide an idea of how the literature is used. This can be used to assess not simply the productivity of a scientist (i.e. number of publications), but their usefulness to others (i.e. number of citations).

Clearly, the Science Citation Index cost some money to put together, and universities paid annually to receive the books, and then the CDs and finally to access the online versions. They were a not for profit organisation, but eventually, they were bought out.

## 13.5   Administrators want metrics

It wasn't just scientists that were interested in the databases of literature. Administrators became interested in the metrics that the databases produced on academics so that they could use them as a measure of performance, or potential in the case of hiring academics. The obsession with simplifying metrics by administrators who don't understand the work that their academics produce, has driven the importance and proliferations of literature databases. However, it is widely acknowledged that the way in which journal metrics are used is unhealthy, and this has led to the San Francisco Declaration on Research Assessment[2].

In the life sciences, there are three prominent databases used: Google Scholar, Web of Science and Scopus (owned by Elsevier). But new players are entering the market all the time, and two of the newest are Dimensions and Microsoft Academic.

## 13.6   Searching by scientist's name

Once you have found a scientist who works within your field and produces influential work that you are keen to read, then you may want to search for all of their other papers to determine whether there is more literature that you might have missed. The databases make such searches easy, but you will need the (correct) spelling of your scientist's name, and also to be aware that they may have used different iterations of their name throughout their career.

### 13.6.1   Overcoming the difficulties of common names

While Dave Thompson could always add his middle initial to specify his work, as time went by there were more and more scientists publishing as D J Thomp-

---

[2]https://sfdora.org/read/%5D%20(known%20as%20DORA)

son. This made it more difficult for administrators and scientists alike. This has begun to be a bigger global problem, such that the databases have begun to use profiles of particular scientists that individuals can 'own' and curate. Each database has its own way of doing this, and it can become rather a chore so that not many scientists have their own profiles, which makes it a problem, particularly for those with common names.

The ORCID number has been erected to provide a common platform for authors to curate, and now some journals and funding authorities won't allow you to submit or apply without one.

## 13.7 Google Scholar, Web of Science or Scopus?

Have you ever wondered why Google Scholar (GS) scores are so inflated compared to other citation databases like Web of Science (WoS) or Scopus? I've noticed that Scopus has better coverage than WoS and that GS is bigger than both (and a lot messier with lots of weird duplicates and poorly entered stuff), but is there anything more to it than that?

Well, it seems that there are some people who have already thought about this, and come up with a good idea of exactly what's different. Martín-Martín et al. (2018) have done a great job of analysing data from some 2.5 million citations. What they found inspired me to write this section, in which I've focused on the results from Life-Sciences. But I encourage you to go read the article for yourself.

I have been known to take the odd peek at my Google Scholar profile over the year and see how the citations are accumulating. I rarely check on WoS or Scopus because it's a bit of a faff getting signed in and doing the search. Plus it looks so much smaller when one is habituated to seeing those double digits in GS! However, I've always been a bit uneasy about citing my GS citation rate, H-index or i10 (among other metrics that they give) as I've never really known what all that extra represents. Something grey and unseemly? Well, it turns out that it's all good stuff, and perhaps GS is the better one to cite as it's a more inclusive index: more inclusive of different document types and different languages.

However, it is worth noting that of the three indices, Google Scholar seems to be the easiest to manipulate. Delgado López-Cózar et al. (2014) show how it is possible to manipulate citations and therefore metrics of individual researchers by depositing false documents into institutional repositories (note that they did this as an experiment, and you may find restrictions on doing this at your own institution are prohibitive or lax). The authors also note that a consistent problem with Google Scholar is the lack of transparency, and therefore verification, of how the bibliometrics are calculated. This will make a difference to what papers are found, as Google Scholar tends to report those with the greatest citations first on the first page of outputs. Hence once a paper is initially boosted by manipulation, it will be more likely to be found and increasingly cited.

To directly compare your citation metrics from different platforms, and also to get more unbiased metrics than these platforms provide, consider using the software Publish or Perish (Harzing, 2007). This also provides a great platform on which to harvest data from Google Scholar into a database, or put it directly into another database, like EndNote, Zotero or Mendeley.

## 13.8    What's the difference between indices?

According to Martín-Martín et al. (2018), the GS score is likely to be inflated – 1.90 for GS/WoS and 1.45 for GS/Scopus. If you deviate from this with a higher score, you can give yourself a pat on the back for having work that's reaching more people in more parts of the world. Google Scholar results reflect a more inclusive citation index (Martin-Martin et al., 2018). While WoS and Scopus aren't exclusively English or journal publications, they are mostly. But that extra third that GS gives you allows you to show the extra scope that your work is getting outside that English journal mainstream. Is your GS score more than a third higher than your WoS or Scopus score? If yes, then your work is having a greater impact elsewhere in the world, and there's nothing wrong with that.

# 14

## Reference Managers

Because published papers play such a fundamental role in your reading and comprehension of a subject, and citations are pivotal in chapter or paper writing, putting these elements together in a good reference manager makes a lot of sense. A reference manager is a piece of software in which you can store and view all of your papers (in pdf format or via links), make notes on them, and use them to make citations and build bibliographies. Unsurprisingly, they become something that you may well use a lot during the course of your PhD, and it is worth thinking carefully about which one you want to use.

There are quite a lot of different software packages out there to help you do this (e.g. Endnote[1], Mendeley[2], Zotero[3] – see more about these here[4]). These packages are a mixed blessing. Firstly, most of them are not free, so unless your university subscribes to their providers you would have to pay for the privilege of using them. Next, all of the ones that I've ever used are very demanding and do not necessarily work well with your word processor of choice. Given all the time I have invested in reference formatting software, I think I might have been better off just copying and pasting references I have at the end of a document and then formatting them by hand. Having said this, I used Zotero as the referencing software for this book. It took very little time to learn how to use, it was painless to enter (most) of the references into the database, and it is free and works with Bookdown[5] and Google Docs.

My advice here is (like software for writing your PhD) to pick the format that your references are going to be stored in carefully. For example, the BibTex format (see below) is a generic format that many of the software packages are capable of importing and exporting. Using a generic referencing format means that you will (in future) be able to change packages with minimum loss. Note that there is a difference between being able to output in BibTex and storage as BibTex files. In the case of the former, you will require a working piece of software in order to output from that format. If your software stores references as BibTex files, you are always able to take this file and use it elsewhere.

---

[1] https://endnote.com/

[2] https://www.mendeley.com/guides/desktop

[3] https://www.zotero.org/

[4] https://subjectguides.library.american.edu/c.php?g=479020&p=3323781

[5] bookdown.org

DOI: 10.1201/9781003212560-16                                                93

**An example of the way in which the generic BibTex format stores references**

```
@article{measey2017counting,
    title = {Counting chirps: acoustic monitoring of cryptic frogs},
    volume = {54},
  copyright = {© 2016 The Authors. Journal of Applied Ecology published
  by John Wiley \& Sons Ltd on behalf of British Ecological Society.},
    issn = {1365-2664},
    shorttitle = {Counting chirps},
    url = {https://besjournals.onlinelibrary.wiley.com/doi/abs/10.1111/
    1365-2664.12810},
    doi = {https://doi.org/10.1111/1365-2664.12810},
    abstract = {Global amphibian declines have resulted in a vital need
    for monitoring programmes that follow population trends....
    acoustically active species.},
    language = {en},
    number = {3},
    urldate = {2021-03-28},
    journal = {Journal of Applied Ecology},
    author = {Measey, G. John and Stevenson, Ben C. and Scott, Tanya and
    Altwegg, Res and Borchers, David L.},
    year = {2017},
    note = {\_eprint: https://besjournals.onlinelibrary.wiley.com/doi/
    pdf/10.1111/1365-2664.12810},
    keywords = {acoustic array, acoustic spatially explicit capture-
    recapture, anurans, call density, non-invasive sampling, population
    monitoring, sensor networks, signal strength, time of arrival,
    triangulation},
    pages = {894--902},
    file = {Full Text PDF:C\:\\Users\\Zotero\\...},
}
```

## 14.1   Other points to consider

When choosing a reference manager, here are some points that you may want to consider:

- Does it integrate with your word processing package?
- Does your university have a licence?
- Do other members of your lab use this software?
    - is there a communal repository?

- Can you use it to read pdf files and make notes on them?
- Does it output into referencing formats that will be useful to you?
    - Or will it allow you to make a custom output format?
- How much time will it take you to learn to use it?
- Is there good support available?

# 15

## What Software Should I Use to Write My PhD?

Thirty years ago, when I started to use computers as an undergraduate, the word processing standard was a piece of software called WordPerfect. Back then it seemed inconceivable that anyone would want to use anything else or that WordPerfect could lose its grip on the market. WordPerfect became a victim of the Microsoft revolution that standardised most PC operating systems (think Windows) and software using Microsoft Office in the mid-1990s. The chances are that today your institution will still have subscriptions to Microsoft Office making it free for you to use. But given that changes in software can happen so fast and appear to change with increasing frequency, should you use Microsoft's Word or another piece of software to write your PhD?

The simple answer is that you should use whatever you, and your advisor, feel most comfortable with. You will face a lot of challenges during your PhD including learning to write with lots of different kinds of software. Each one you learn to use will require a steep learning curve, and that will require your time. If there's any good time to start with a new word processing package it will be before you start your PhD. If you do decide to go with a new writing package then my suggestion would be to choose open source software like (RMarkdown Xie et al., 2018). More on these later.

Another important consideration is whether or not your chosen word processing package will allow seemless integration with your reference manager. Your lab may have a communal repository of references, and so this may then constrain what you use.

If you feel comfortable with a word processing package with a nice GUI, like Microsoft Word, then go ahead and use it. These certainly have advantages like being able to get a synonym at the click of a mouse (Figure 9.1). My best advice is that you do not save your files in their default ".docx" format. Instead, make sure that all of your files are saved as .rtf, .txt or .html. This should make little difference to your opening and editing but will mean that all of your texts are forward compatible with other kinds of software. It will reduce your file size and reduce the chances of file corruption: i.e. when part of the file malfunctions and you lose the contents. That's right, in addition to

DOI: 10.1201/9781003212560-17

loss of data by back up deficiencies (see Part III), you also have the risk of having corrupt files that will also mean you lose all of the contents. Even if you always use Word and never use any other file format, do yourself a favour and save each version as a .rtf or .txt file as well. Just like a back up, this means that you won't have lost all of the text if a file goes corrupt. And yes, I do know people who have multiple backup versions of corrupt files: it's happened to me.

Keeping files in a generic (rather than software manufacturer's) format will allow easy sharing of these files, and access to them in future. It might seem inconceivable now that anyone would not be able to read a .docx file, but in 10 years, you may be so used to the latest format that after another five years it becomes very difficult to retrieve information from a .docx file without paying someone to retrieve your data.

The real problem with word processing packages like Microsoft Word is when files get very large, as they tend to do during a PhD. Once you start putting chapters together, especially if they include embedded images tables and the like you will find that many word processors start misbehaving. This can be especially frustrating in the closing phases of your PhD when you're trying to submit it. Especially if there's a tight deadline. However, this problem is easily overcome by chunking your thesis into documents equivalent to chapters and then merging PDFs from each chapter.

## 15.1 Alternatives to conventional word processors

Many academics are turning to LaTeX in order to write their manuscripts as this gives a lot of possibilities to insert links, formulae, references, images, etc., in a very simple way. Once you are ready to print a LaTeX document it can be shared very easily as a PDF.

Be aware that the problems may come not with what you want but with how your advisor and or co-authors are comfortable. I find that commenting on a PDF takes more time than doing the same on a Word document. I don't mind working with LaTeX but I know some advisors are very averse to having to use this platform. Obviously, in order to get the feedback from your advisor you are going to need to use a platform that they are also comfortable with. So please consult them on what they are happy to use. You may also need to instruct them about how to make comments. For example, track changes is a function that a lot of people like to use but it doesn't translate back into other platforms seamlessly.

As with many things in this book I urge you to keep an open mind with respect to software. For example, if your lab is progressive and already experimenting with new and flexible software platforms then I suggest that you learn with them. The future of computing is certainly likely to be open-source software, and so the sooner that you become more familiar with this, the easier you will find it to adapt in future.

Another consideration is how your word processing package is going to work with referencing and citations. Again, your lab may already use some software that you want to integrate with your word processor.

If you are feeling adventurous, try writing in R Markdown, using RStudio as a text editor GUI (Graphical User Interface) and the Bookdown[1] platform to publish your chapters into a single thesis file in virtually any format. There is also great support for both of these formats with free guidebooks about how to use them (Xie, 2016; Xie et al., 2018). Given that I wrote this book in R Markdown with Bookdown, I feel that I should point out (albeit briefly) the advantages to this platform:

- Reproducibility
- Dynamic Content (include statistical code and graphics inside your thesis)
- Focus on content (not formatting)
- Numerous output types (including .docx, .html and .pdf)
- Independent reference manager (via .bib file) and style files (via .csl file)
- Free software and community support

Importantly, what you want to try to avoid is conducting a major switch between software right at the end of your PhD. Try making such changes when you have more time.

## 15.2 Naming versions of files

Good file management is an important aspect of a PhD. I like my students to give each chapter of their thesis a nickname or short running title that we both agree on to name the files. An example might be "Swimming_performance_V1.rtf". The reason why I don't like "Chapter_1.rtf" as a filename is that every student has a chapter 1, so I can't distinguish between them. In addition, I'm not good at remembering which chapter is chapter 1 and which is chapter 2. Sometimes, towards the end of a thesis, the chapters get their orders switched around, causing even more confusion. Notice that I don't like having spaces in filenames. This is a hangover from old days of MS-DOS but is consistently useful today when using R and other coding software.

---

[1] https://bookdown.org/

However, some software platforms react oddly with underscore (_), and so you might need to work with a hyphen instead of a space (-).

Once I've commented on a file and sent it back in an email or uploaded it to their shared folder, I'll add my initials Swimming_performance_V1_JM.rtf. Each time I comment on this version, I will add a number to my initials and the student adds one back, so eventually, we might come to: Swimming_performance_V1_JM3_MM2.rtf. After three rounds, we would normally send a clean V2 out to co-authors and once their comments are back in we should be all set.

While you and your advisor can adopt any naming convention you like, you and they need to instantly be able to distinguish between a new version and one that they have already commented on by the file name. No advisor wants to spend time commenting on a chapter and then be told that it was an old version.

## 15.3   How many versions of each file should you keep?

I would suggest keeping at least one older version of each file to minimise the possibility of losing the content through corrupt files, but you should also be keeping those .rtf or .txt files as well, just in case. Because of my early experience with file corruption, I started to keep all versions of files. This is unnecessary, and I've never needed to go back to anything but the last version. Nevertheless, text files take up very little space and it's not a problem to keep more versions if you want to. More on this in the chapter on data management.

# 16

## What to Do When Faced With a Paywall?

Unless you are at one of the most prestigious institutions, you will likely encounter the paywall from time to time. This is when you cannot access the content of a paper that you want to read, but instead are asked by the publisher to pay for access. There are many legitimate options that you have to access such content. Ultimately however, you can always ask for an interlibrary loan. This is last on the list as someone will have to pay for it and it will take some time. Your library may have an allocation that they can use before the bill comes to your advisor's desk.

## 16.1 Go to your library

Unsurprisingly, this should be your first port of call. It's probable that your library has an electronic portal that you can use to access all of their subscriptions. But don't forget that they have stacks with old copies and you might need to physically enter the building. Try it. They are great places!

## 16.2 Look for links from Google Scholar

Google Scholar has lots of links (on the right) labelled with "[pdf]," and these are often integrated with your library. If there's nothing at first, click on the "All 9 versions" link (below and to the right). This may reveal that there is more than one version with a pdf, and you might have access to one of the others. Check for a link to the author on google scholar, and then look for a link to their website.

DOI: 10.1201/9781003212560-18

## 16.3   Google the title in quotes

Sometimes, Google scholar won't show a pdf, but you can Google the title "in quotes" and get back a direct link to the pdf. Although rare, it's worth a try before moving on.

## 16.4   Go to the author websites and look for links

Authors often have official university websites, and sometimes their own home curated sites that may have pdfs of all their papers. Use the clues at your disposal, including a general search engine, and follow the trail.

## 16.5   Go to the institutional repository of the author

Most university libraries curate an institutional repository of pdfs of published final manuscripts accepted for publication, so-called Green Open Access copies. These are identical to the printed versions save for the typesetting and proof reading that is done after this accepted copy.

## 16.6   Try an Open Source OA tool

There are an increasing number of Open Source tools that aim to legally find articles for you. A good example is Unpaywall which has more than 30 million articles and this OA tool is growing all the time. Another tool, OA Button works as a browser extension which will do the searching for you (i.e. most of what is above), also returning a legal OA version of the text you are looking for. These tools are a great way of getting access to OA sources, but note that it will only work if you and others label your Green OA work with the publishers' DOI.

## 16.7 Write to the author and request a pdf

Some authors will respond very quickly to reprint requests, while you won't hear from others for some time, or at all. However, this was the traditional way of obtaining a copy that you didn't have access to, and is definitely worth trying. You should be able to get the email address of the corresponding author from the journal website. Sometimes these are hidden and you need to jump through hoops to find them. You could also search for the email addresses of any co-authors, and also write to them.

## 16.8 Go to an academic social media site

Many academics also have accounts on academic social media sites (e.g. Academia.edu; ResearchGate.net). You will need to join the network (using an email address from an academic institution) and have your own account, but these house a lot of pdfs that you can't get elsewhere.

## 16.9 Black open access

There are sites that host papers that infringe the copyright holders' rights, often termed **Black OA** sites. These sites are blocked by some institutions and are inaccessible from some countries.

## 16.10 Use your contacts at other universities

The chances are that you have friends at other universities where the subscription you are looking for is not behind a paywall. If not, then ask your advisor or lab mates, or post it on your lab's (electronic) notice board.

## 16.11 Go back to your librarian and ask them for help

Librarians have contacts at other libraries. Some have had careers at other institutions or have met people at conferences who'll be prepared to help them. Librarians are great people and well worth getting to know. Their job is to help you get stuff, so it's always worth engaging with them.

## 16.12 Get an interlibrary loan

This comes last on my list as it costs money. If what you are after is a book, then you might end up going here very quickly as many older books are not available online. If it is a paper, it will likely be copied or scanned into a pdf where it occurs. If it is a physical loan, the interlibrary loan will be time limited. When filling out the form, make sure you know where the loan will be billed to before you submit it. And don't come here first, the chances are that you can get it somewhere in the list above much faster and cheaper. Note that there are Open Source tools now available for your librarian that can make interlibrary loans much cheaper, for example InstantILL.

# 17

# Scientific Names and Taxonomic Authorities

There are quite a few conventions that it's worth being aware of when it comes to using names. When talking about common names, it is a bit of a quagmire as there are few standards that are followed by all journals. Scientific names can also be in great (taxonomic) flux, but at least there is definitive help. Here I will outline my interpretation of what's what in using names.

## 17.1  Scientific names

There is only one valid scientific name for a species. This is expressed in a binomial: two names, first the genus (with an initial capital letter) and then the species name (lower case). This name is italicised by convention (or, in some cases, underlined). Note that no other taxonomic names (e.g. Family, Order, Class, Phylum) are written with italics (although they all get capital first letters), only the genus and species are italicised. The single name is a hard rule as taxonomy can't survive in any other way. The most recently published version is the one to use, but happily with many vertebrates (mammals[1], birds[2], fish[3], reptiles[4] and amphibians[5]) there are well-curated databases to which you can refer for the latest taxonomic treatments. Also, see WoRMS[6] for marine species. My suggestion is that you defer to these (with citation) and you won't go far wrong. If you do this early on in your thesis, remember that you may need to update it later.

### 17.1.1  Consider this example

When my former student started his PhD in 2013, he worked on three species, and over time these changed so that at various points in his thesis they had

---

[1] https://www.mammaldiversity.org/

[2] https://avibase.bsc-eoc.org/avibase.jsp?lang=EN

[3] https://www.fishbase.se/search.php

[4] http://www.reptile-database.org/

[5] http://research.amnh.org/vz/herpetology/amphibia/

[6] http://www.marinespecies.org/

DOI: 10.1201/9781003212560-19

a range of names. You can see the combinations he had to contend with in Table 17.1.

**TABLE 17.1: Taxonomy is dynamic which means that species names change over time.** The way in which genus and species names can change in only a period of five years.

| *Amietophrynus rangeri* | *Amietia angolensis* | *Xenopus laevis* |
| *Amietophrynus rangeri* | *Amietia quecketti* | *Xenopus laevis* |
| *Sclerophrys capensis*[7] | *Amietia delalandii*[8] | *Xenopus laevis*[9] |

The taxonomy of two species changed such that he had to constantly revise the names in his thesis. Luckily, he could simply conduct a 'replace all', but it did become quite confusing. There's a chance that they've changed again, so I've provided hyperlinks for you to check out what the current name is. The lesson is that if you want to avoid this, work on *Xenopus laevis* or conduct your PhD in as short a time as possible!

In taxonomic papers and certain journals with a taxonomic focus, there is an insistence that you cite the 'taxonomic authority' after the species name when used for the first time. This is essentially a citation in which you acknowledge the original description that accompanies the name. Let's take the above examples again:

*Sclerophrys capensis* Tschudi, 1838
*Amietia delalandii* (Duméril and Bibron, 1841)
*Xenopus laevis* (Daudin, 1802)

---

[7]https://amphibiansoftheworld.amnh.org/Amphibia/Anura/Bufonidae/Sclerophrys/Sclerophrys-capensis
[8]https://amphibiansoftheworld.amnh.org/Amphibia/Anura/Pyxicephalidae/Cacosterninae/Amietia/Amietia-delalandii
[9]https://amphibiansoftheworld.amnh.org/Amphibia/Anura/Pipidae/Xenopus/Xenopus-laevis

It might seem that there's a mistake above because Tschudi, 1838 isn't in brackets. This is actually deliberate. Tschudi, 1838 isn't in brackets because the genus name hasn't changed since the description by Tschudi 1838 placed *S. capensis* in *Sclerophrys* in 1838. But it is clear in the table above that the genus name did change to *Amietophrynus*, and it was recognised under a different species name (*A. rangeri*). Unlikely as it may seem, the original description of the species was by Tschudi in 1838, but this was long forgotten (or ignored because the type specimen was so poorly preserved), and the species was regularly referred to as *Bufo rangeri*. When the genus *Bufo* was broken up into lots of newly named genera by Frost et al (2006), Tschudi's description was once again disregarded (Ohler and Dubois, 2016). However, in 2016 new work on the type specimen revealed that it was the same as *Amietophrynus rangeri*, but because Tschudi's description was older, the taxonomy deferred to the rule of precedence (or primacy). If you want to read more about it, you can look at all the changes on the amphibian taxonomic database.

To clarify, the use of brackets around the taxonomic authority only happens when the species is no longer in the genus in which it was originally placed by the person who described the species. Daudin (1802) originally described *Bufo laevis*, and Duméril and Bibron (1841) originally described *Rana delalandii*. Hence you will see that both of these authors are within brackets after the names of their respective species.

The generic name of a Latin binomial is often abbreviated to the first letter (e.g. *X. laevis*). This should only be done once it is well established in your writing what the full name is. Or, in other words, you should refer to the species in full on the first mention and then switch to the abbreviation. Because sentences shouldn't start with an abbreviation, you might need to relax this from time to time. You also need to treat the abstract as a stand-alone document in this respect. You might find that you have two species, which belong to different genera that share the same first letter. In this case, most authors use a different abbreviation: for example, *Ap. rangeri* and *At. angolensis*. This is exasperated when both species, names are the same.

I have seen some comments that suggest that once a generic name is abbreviated in a paper, it shouldn't be written in full again. In my opinion, this is rather a silly idea. It is up to you (the writer) to avoid ambiguity. If there is no ambiguity, then you may write the genus and species in full to increase the readability of your piece.

You should not start a sentence with an abbreviated genus. *X. laevis* would not be good because your first word of the sentence is abbreviated (*X.* ). *Xenopus laevis* is a perfectly admissible way to start a sentence.

Lastly, there is a convention that when referring to the generic name alone you should always mention that it is a genus. So you shouldn't write about those *Bufo* that hop around the house, but instead about those members **of the genus** *Bufo* that hop around the house.

---

## 17.2 Different codes for the nomenclature of different taxa

Nomenclatural codes exist to ensure stability in names. At first glance, this might seem like quite a trivial (academic) matter. But think about what happens when a particular species is named in law. That name could (in theory) be changed by some unscrupulous biologist to open up a legal loophole, or at least some ambiguity. Once you are aware what can go wrong with taxonomy and the implications that it can have on many lives, you will understand the need for stability. Happily, there are associations of very serious people who guard the stability of taxonomy through codes. It is important to realise that there are codes governing the naming of different levels of taxa.

- Viruses – International Code of Virus Classification and Nomenclature[10]
- Bacteria & Archaea – International Code of Nomenclature of Prokaryotes[11]
- Algae, Fungi & Plants – International Code of Nomenclature for algae, fungi, and plants[12]
- Animals – International Code of Zoological Nomenclature[13]

There are many peculiarities for each taxon, and if your PhD involves taxonomy explicitly you should consult the specific code. Otherwise, it's unlikely that you'll need to be aware of the contents of each or any code.

### 17.2.1 Peculiarities of plant taxonomy

Plant taxonomy is considered the taxonomic mother discipline as the work of Linneaus in 1753 was the first botanical work that consistently used Linneus' binomial (two names: Genus species) nomenclatural system (Linnaeus, 1753) which pre-dated the more inclusive (and better known) landmark publication in 1758 (Linnaeus, 1758). Until the Botanical Code was revised in 2012, all new botanical diagnoses had to be written in Latin. In addition to the Latin binomial name (Genus and species), plants (and some other taxa) often require further descriptors or **ranks** that denote subspecies, varieties, subvarieties, etc.

---

[10]https://talk.ictvonline.org/information/w/ictv-information/383/ictv-code
[11]https://www.the-icsp.org/bacterial-code
[12]https://www.iapt-taxon.org/nomen/main.php
[13]https://www.iczn.org/the-code/the-international-code-of-zoological-nomenclature/

I have included a (non-exhaustive) list of these in Table 17.2, together with some examples.

**TABLE 17.2: Plants often need more information than the simple Latin binomial.** This table provides the required abbreviations that you will need to add in order to properly describe the rank of your plant taxa. Note that the taxonomic authority (last column) can include two names with the first in parentheses if the second changed the placement of this taxon.

| Rank | Abbreviation | Example | Authority |
|------|-------------|---------|-----------|
| Subfamily | subf. | Poaceae subf. Pooideae | Benth. |
| Subgenus | subg. | *Metrosideros* subg. *Metrosideros* (Myrtaceae) | |
| Section | sect. | *Senecio* sect. *Senecio* (Asteraceae) | |
| Subsection | subsect. | *Stachys* subsect. *Swainsonianeae* (Labiatae) | |
| Series | ser. | *Mimosa* ser. *Paucifoliolatae* | Benth. |
| Subseries | subser. | *Palicourea* subser. *subcymosae* | (Müll.Arg.) C.M.Taylor |
| Subspecies | subsp. | *Trichophorum cespitosum* subsp. *germanicum* | (Palla) Hegi |
| Variety | var. | *Closterium acerosum* var. *elongatum* | Brebisson |
| Subvariety | subvar. | *Hieracium vulgatum* subvar. *oblongatum* | Sudre |
| Form | f. | *Prunus cerasifera* f. *atropurpurea* | Diffel |

Unlike animal authorities, for plants, if the combination of Genus and species has changed, both the authority for the original genus placement, and the author responsible for the new combination are given after the species name (with the former in brackets see Table 17.2). In addition, for plants no year is given, and the name of the authority is usually abbreviated. Because Linneaus named so many plant species in his landmark publications, instead of writing out the full taxonomic authority after the species names, botanists abbreviate this to "L.". Like Linneaus, other prolific plant taxonomists can have their names abbreviated, but plant taxonomists became so numerous that there needed to be some standardisation of names and their abbreviations. A database of these **Standard Forms** can be found at the International Plant Names Index database[14]. Note that Standard Forms in plant taxonomy can include initials.

---

[14]https://www.kew.org/data/authors.html

Algae sometimes exist in association with other taxa (such as lichens which are an association between an algae or cyanobacteria and one or more fungi. For these associations, there is a specific code for their nomenclature: International Code of Phytosociological Nomenclature (Theurillat et al., 2021).

There is also a code for cultivated plants, and if you study any kind of agronomical system, you will be familiar with not just the normal binomial name, but also a specific cultivar or variety (*var.*) that follows the binomial: The International Code of Nomenclature for Cultivated Plants[15].

### 17.2.2   Other times to use italics

In general, italics are used when names are derived from Latin or Greek. A number of subdisciplines have international codes for how the names should be written: Anatomy, Histology, Embryology and Neuroanatomy. There is often confusion about these correct terms, but luckily four major publications are here to help you. These are maintained by the international groups who specialise in standardising the nomenclature, and they are republished from time to time (so look out for updates). I provide directions to these citations should your research require them:

- Terminologia Anatomica. (Federative Committee on Anatomical Terminology, 1998)
- Terminologia Histologica. (Federative Committee on Anatomical Terminology, 2008)
- Terminologia Embryologica. (Federative Committee on Anatomical Terminology, 2013)
- Terminologia Neuroanatomica. (Federative Committee on Anatomical Terminology, 2017)

### 17.3   Common names

Common names have a different set of problems to those that we've seen for scientific names. These are largely focused on: what name to use, and whether or not to capitalise the common name.

**What name to use** – This is tricky. My suggestion would be to take the most commonly used common name, as that's likely to be the one that most people

---

[15]https://www.ishs.org/scripta-horticulturae/international-code-nomenclature-cultivated-plants-ninth-edition

will recognise. However, it's not always that simple, and because there's no 'correct' common name, you have some leeway to choose the one you prefer.

Let's take an example: The African Clawed Frog is a very widely used name for *Xenopus laevis* which is used all over the world as the model amphibian. However, African Clawed Frogs is also the commonly accepted name for the entire genus, *Xenopus*. Thus, shouldn't we avoid using this for the species *X. laevis*? So, let's look at the alternatives. Frost (2020) lists 13 common names for this species. Names such as Platanna and Common Platanna are difficult to use in scientific papers as they are only regularly used in South Africa. Clawed Toad is not appropriate, as it is not a toad (reserved for members of the family Bufonidae), a confusion that dates right back to Daudin's 1802 description. Clawed Frog has the same problem as African Clawed Frog, and other variants such as Upland Clawed Frog and Smooth Clawed Frog also seem inappropriate as this species is neither confined to uplands, nor the only member of the genus which is smooth (although 'laevis', the species name, does mean 'smooth').

There has been a recent movement to formulate common names into genera (or even families) such that the specific common name takes on a specific epithet of the generic or familial common name. In a world of stable taxonomy (such as that enjoyed by ornithologists), I can understand that this is possible. But imagine the requirements in common name changes for frogs in Table 17.1 over only a few years. Surely the useful thing about common names is that they are used commonly by normal people who don't worry about taxonomy? If we insist on common names constantly changing, don't we risk alienating the public?

I eventually decided to use 'African Clawed Frog' as this is the most widely recognised common name for this species. It does become an issue when writing about *X. laevis* and other *Xenopus* species (as we have), but we've managed to resolve this without too much difficulty.

---

## 17.4 Should common names have capitals?

**Warning: lots of people get very upset about this point and there are two views**

### 17.4.1 View 1 – Yes

Common names are proper nouns and proper nouns start with capitals. The Western Leopard Toad is not any toad, and the African Clawed Frog isn't any frog. The Cape platanna is any platanna living in the Cape, but the Cape

Platanna is *Xenopus gilli*! In this last example, you can see how capitalisation reduces ambiguity and differentiates between adjectives and nouns.

## 17.4.2   View 2 – No

Common names are not formal names (like scientific names) so why does it matter? Many journals do not capitalise common names and insist on lower case unless they are proper nouns, like African in African clawed frogs. This seems to have arisen from editors who dislike superfluous capitalisation.

On balance, I'd say that it's likely that your usage will be dictated by the policies of your faculty, advisor or journal.

# 18

## *Writing Style*

**Did you ever feel that reading a paper was like trying to solve a puzzle?**

During my PhD I read a lot. My study species was the subject of tens of thousands of papers, and I was convinced that I'd find what I'd needed somewhere in those musty old reprints and heavy volumes that came from the library shelves. It was easy to believe that the authors of those papers had set great puzzles for me to try to understand somewhere in their content. The realisation that this is not the best form of communication came as a personal epiphany some years ago.

---

"The goal of good writing is straightforward: to make your reader's job as easy as possible."

Kevin W. Plaxco (2010)

---

This great quote from a paper by Plaxco (2010) who cuts to the heart of why it's essential to write well. As an editor, I once experienced an author who thought the opposite, and for me, this was the epiphany. The submitted manuscript was an impenetrable mess. I could tell that there was good work in there, but as an editor, I felt that the information had been made so obscure by the authors that my readers were unlikely to get much out of it. The author was a colleague, and so I decided to phone him and chat about the need for much greater clarity in his manuscript. Experience with email has taught me that they can often be taken in the wrong way; usually in the worst possible way.

The response I got surprised me. The author recognised that his text was dense and was unapologetic. "Let the reader work to make sense of my data," he said. My epiphany came not because this was a totally alien concept but because during my PhD days, this was exactly how I had thought scientific writing should be done. I had spent so many hours slogging through dense and dreary papers written by well-respected figures. The reward came when

DOI: 10.1201/9781003212560-20

I finally understood what it was that they were trying to communicate. This felt so great that I believed the puzzle they had set me was what I should set my own readers. Happily, my advisor didn't feel the same way, but for years I continued to believe that a paper should be a decent puzzle for readers to crack. This might seem a crazy concept, but it's not an uncommon problem. For a great take on this in Chemistry, see this article by Murray (2011).

The preponderance of impenetrable guff has led to a general perception that "academics stink at writing" – as Steven Pinker put it (Pinker, 2014). But, I think that Pinker (2014) is outdated. Certainly, as Plaxco (2010) has it, there are a lot of ways that academics can improve their writing for each other. But the belief that all academic writing is going to be accessible to the general public is far fetched. This is because there are a lot of terms when writing journal articles that are there for precision and brevity. These terms will exclude a larger audience, especially when they get so dense that looking them up is simply too arduous. What Plaxco (2010) urges us to do is to make the rest of the text as accessible and easy as possible.

Any good writing is an artform, and that includes good academic writing. You can't expect to be the best until you've spent some time at it. If you consider a portfolio of your completed written work as 'your experience', you will get some insight into exactly how small this is, even once you've finished your PhD.

There are still authors out there who attempt to set puzzles for their readers, but they aren't in the mainstream any more. Instead, biological sciences have some inspiring writers, and many more are taught how to inspire future generations towards communicating great science. That's not to say that there are no puzzles left. Much of what we do requires great puzzle-solving abilities. However, let's keep the puzzles away from communicating with our audience, make their job as reader easy or even pleasurable, and they'll keep coming back for more.

## 18.1   Develop your writing style

During the course of your PhD you should improve your writing style towards proficient communication with your target audience. This will involve improving both your technical writing and accessibility writing skills. Although the bottom left portion of Figure 18.1 is labelled 'Poor communication', we might more productively think of it as being the area of the 'first draft'. You have to start somewhere. If your current style is too 'hard' and technical (top left), then you will need to increase your accessibility. If, on the other hand, you are used to a more popular writing style (bottom right), you will need to increase your technical ability. See Dunleavy (2003) for more insight on achieving your requisite writing style.

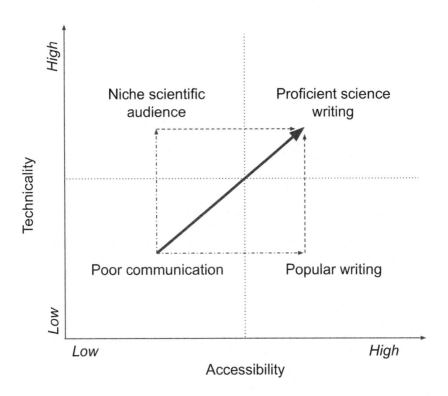

**FIGURE 18.1: Pitching your writing style will take some practice to reach the desired pitch.** Wherever your starting point in this technicality and accessibility square, your aim is to move towards the top right corner. Adapted from Dunleavy (2003: Fig. 5.1).

## 18.2 Don't bamboozle

It's easy to use jargon. The whole point of jargon is to convey a (usually complex) idea in a short amount of space. Using a word (or two) instead of using several sentences clearly has some advantages. However, there is such a thing as too much jargon. Simply put, it's unnecessary to use jargon when you can use plain English in the same amount of space. My old advisor at Liverpool University, the late, great Brian Moss, shared the following example of too much jargon by Bump et al. (2009), when plain English would have been much shorter. Do yourself a favor and read the abstract of Bump et al. (2009). The fact that it gave Moss a chance to write about moose wasn't lost on anyone (see Figure 18.2)!

> Dear Sir, I have just attempted to read a paper in one of the Society's Journals about the transfer of nitrogen from lakes to land by the excretion and death of moose. The Summary should have read:
>
> *'Moose feed on plants in rivers and lakes and excrete and die on land, thus transferring significant amounts of nitrogen from the water to the land. The pattern of transfer is not random and depends on where moose feed and where wolves kill them.'*
>
> Instead I had to wade through nearly five times as many words, including such horrors as 'associated spatio-temporal patterns of resource flow across aquatic-terrestrial boundaries', 'cross-habitat resource flux, 'significantly clustered at multiple scales' and 'faunal mediated resource transfer. In neither the title nor key-words do any of 'nitrogen', 'moose' or 'wolf' (or their scientific names) occur. The main text was just as bad.
>
> Do we not have some duty to protect the English language and the sanity of the readership against this sort of thing? It is very common in the Society's journals. Editors are there to edit. Why are they not editing any more?
>
> Yours sincerely, Brian Moss (ex-editor, Journal of Ecology)

**FIGURE 18.2: Bamboozled Moss takes on the moose.** Brian was so unhappy after reading Bump et al. (2009) that he felt compelled to write a letter to the British Ecological Society Bulletin.

Whatever your writing style, you need to be conscious of your ultimate aim in writing: to communicate with your reader. Making your text as easy to read as possible (as Plaxco, 2010, put it) isn't easy, but it will be rewarding. The reward you will get is having willing readers understand the message that you want to convey. This is something that will stand you in good stead no matter what career you have after your PhD.

# 19

## Retaining Your Own Voice When Writing

Should one use 'I' or 'we' when writing a scientific paper? Although there was a tradition not to use first-person pronouns when writing scientific articles, this has fallen by the wayside in recent years because the use of I or we makes writing simpler for writers and easier to understand for readers. Use 'I' if you are the sole author (or in your thesis) and 'we' for two or more authors.

You will regularly see 'I' or 'we' being used in the last paragraph of the introduction and the first paragraph of the discussion. It's also quite common to use this in the methods section. The reason why these sections have 'I' or 'we' is quite obvious, as it allows you to place your own voice on your actions and decisions. After all, it is your aim and hypothesis for your study, so you should own it. Using first-person pronouns also allows you to be more concise, and there are some great examples given by Kurt Vonnegut (1980) where you can see what happens when you don't use them.

The first use of 'I' or 'we' at the end of the introduction is powerful. Up to that point, the reader has read four or five paragraphs of logical arguments outlining the background information that makes up the reasoning for your study. Using the first-person pronoun now makes the reader sit up as they have you communicating directly with them on what you aimed to do. But that power is lost with repetition. So if you were to continuously use it throughout the methodology, the reader may get tired of reading a string of statements stating that you did this, then you did that, and so on. The power of using 'I' or 'we' is to use it sparingly to boost clarity.

### 19.1 Does writing 'I' or 'we' help people to understand who you are?

Actually, all of your writing will do this. As long as you are using your own words (and not plagiarising), your writing is likely to be unique to you and hence recognisable as you. It never ceases to amaze me that it is possible to convey the same information in so many different ways. Your way will likely

DOI: 10.1201/9781003212560-21

change with time and experience, but remain yours throughout your writing life. If you keep writing, people will be able to recognise your distinct style.

## 19.2   The Vonnegut advice

Kurt Vonnegut (1980), infamous novelist, provided some advice for writers. First, he made an impassioned plea for technical writers (yes, you) to use your own writing style. The easiest way, Vonnegut argues, is to write about a subject that you care about. He's totally correct on this, and I'm hoping that after all your research you care passionately about your subject. Next, not to ramble, but to keep your writing simple.

What is most impressive about Vonnegut's advice is that he argues from the readers' standpoint. His parting advice:

> "Our audience requires us to be sympathetic and patient teachers, ever willing to simplify and clarify - whereas we would rather soar high above the crowd, singing like nightingales."
>
> — Kurt Vonnegut (1980)

Having said that you should write about subjects that you feel passionately about, this does not mean that you should write with unsubstantiated passion. For example, it makes no sense to write: "The threats to this spectacular species surpass any mechanism known to humankind" ...unless you have substantive citations or results to back this up. Yes, it's passionate, but that should be balanced by the need to keep what you write factual.

# 20

## *Writing Concisely*

In this book, I have put an emphasis on a writing formula in order to get started writing. Many students have a problem in knowing where and how to get started with what seems to be a daunting prospect: writing a proposal or writing the entire thesis. I still feel that getting started quickly and efficiently with some confidence is one of your most important first steps. But once you've made it through your first draft, you'll begin to find that you need to start refining your writing. That is saying what you need to say in less words and space. A bit like constantly rewriting your Tweets to get the same information into 140 characters (yes, it's been increased to 280 but that's not as concise).

In an ideal world, what goes into your thesis will be of the same quality as an accepted manuscript. The reality is that the thesis is often more wordy, with the result that reviewers and/or editors will ask you to be more concise. How can you achieve this?

Scott Hotaling has written a great essay with 10 top rules to help you be concise with your own writing. The essay is Open Access, and there's no need for me to repeat Scott's (2020) words. However, I am going to write some comments of my own on his rules.

## 20.1   Rule 1. Take writing seriously

It is likely, that like me, you chose to do science at school instead of useful subjects such as English and/or other languages. It might have also occurred to you that it was a really bad idea not to have done more earnest work in languages now that you are expected to write like a professional. Certainly, if you did pay more attention in languages in school, you may not be suffering with the rest of us now. Personally, I never received tuition in the grammatical construction of English (you may know this already). Thus I'm left in the dark most of the time about exactly what is wrong. Instead, I simply try to rework a sentence into what I know is correct (and I often don't achieve that). The lesson is that should you come across an opportunity to learn more about writing, do take it. Otherwise, try to practice writing concisely in all that you do.

Use writing professional emails (to your advisor, co-authors or collaborators) as an opportunity to write concise messages that communicate efficiently and precisely the aim of your message. As a PhD student, you should consider yourself a professional and these communications to colleagues will be received with more import if you can write concisely. A personal plea from me would be to ask that you don't start your email by enquiring after the health of your recipients. The fashion to do this has turned into a surge in recent years (the pandemic notwithstanding), and the reality is that these are simply superfluous and gratuitous words. You certainly don't expect them to write back starting with a paragraph about their health. Save this kind of correspondence for your friends and relatives, who may well appreciate you enquiring about their health. When your advisor, co-authors or collaborators become friends, you can move to a less formal style, but I'd argue that you always want to keep your email communication concise.

Although we talk about writing science here, the grammatical rules of writing (in English) are fairly universal (with some local exceptions), so do practice your writing. Maybe write a letter (or long email) instead of making a call. It'll also help you to read critically. If you are a keen reader, then you could do worse than reading some Kurt Vonnegut (1980).

## 20.2   Rule 2: Identify and stick to your message

It's important to avoid distracting your reader (see Part II). Your aim is to be thorough, which will mean including all relevant information but don't allow your writing to sideline them into taking the wrong direction. You need to stick to your narrative like a highway. The highway analogy is useful as although on a highway you will see signs to other places that you can point out to the reader, you shouldn't turn off the highway. The best of highways will also be free of traffic jams (think long, complex sentences here) and toll free (no paywall).

This will mean that you may need to delete some of what you've written, and that can be hard (especially when they are the best bits). You can always keep a file with all the best bits that you've never used. Maybe one day you'll use them. Or maybe one day you'll see that they weren't quite as good as you thought they were.

## 20.3   Rule 3: Get to the point

If the highway is 'your message' then getting to the point is the big sign that states you've arrived at your destination.

Some of the best papers I've read manage to encapsulate the whole point in the first sentence, or better still, in the title. Our formula has you getting to the main question in the last paragraph of the introduction, but you should have already 'got to the point' in the first paragraph - i.e. the point that is the bigger picture.

## 20.4   Rule 4: Keep your methods and results contained

You can find a guide to writing your methods in Part III. This rule is about not allowing the methods to creep out of the methods section and into the results, or even the discussion. Similarly with the Results. There are some specific times when this is permitted (such as a *post hoc* test), but generally, you shouldn't expect to do this.

## 20.5   Rule 5: Do not repeat yourself (too often)

Redundancy is often rife in proposals, theses and manuscripts. If you've produced a table with all of the results, then they don't need to be in the text. The same with a figure, especially Figure 1 which is often a descriptive map or diagram of apparatus. Have it once, but you don't need it twice.

Copy and pasting are very easy, and a good way of suddenly producing vast quantities of text (Figure 22.1). But the repeated text is quickly recognised by the reader and appears very boring and cumbersome. This is likely to happen in the Methods and Results sections. If you find yourself deciding that you can simply cut and paste this paragraph while changing the variable names and the numbers, then you are wrong. Don't ever do it.

Also, please don't cut and paste sections of methods from one part of your thesis to another. Just don't do it.

People also have a tendency to build the abstract by copying and pasting text from the main sections. Don't do it. The reader will quickly see the repetitions and become bored. Similarly, the conclusion/summary section is also often copied from the lines above. If it's not worth writing again, then it wasn't worth writing before. Make it fresh, and keep it interesting!

## 20.6 Rule 6: Avoid unnecessary or inefficient 'lead-ins'

This rule relates to keeping your text concise, specifically at the start of a sentence. You may be tempted to reiterate your point, but this is not always needed. I'd say that you need some practice and a critical eye to spot these kinds of errors. Having your advisor help or someone who has edited a lot of text. Scott's got some great examples (Hotaling, 2020).

## 20.7 Rule 7: Use first-person, active voice

For some reason, most students avoid this at all cost. There appears to be the idea that saying 'I' or 'we' isn't correct for scientific writing. In fact, it can be the easiest way to avoid the passive voice. There are other reasons why using the first person is helpful.

## 20.8 Rule 8: Remove unnecessary words

This is really getting down to the nitty-gritty. It's hard to do this yourself. It's much easier for someone else to show you. Typically I only do this level of word-smithing for abstracts or when the imposed word limit means that you really need to remove excess words. However, Scott is correct that if you can learn to do this yourself, it will improve your writing.

Personally, I like to slip in the odd 'as well as' instead of 'and', just to ring the changes. Scott would remove them, and you can see where he'd edit out other examples. Do we really have to be so hardline? I'd say that there are times when it helps.

## 20.9 Rule 9: Simplify your language

This is always a good idea (see Figure 18.2). Making three short sentences instead of one very long one is much better. It also helps you avoid complicated grammatical clauses.

## 20.10 Rule 10: Seek and embrace feedback

I spend a lot of my time reading and commenting on the writing of my students. Every comment is made to improve the document. I do get mad when I find that this help is ignored without an indication why. Occasionally, my suggestions or comments are simply deleted. Certain students become very hard work when they don't explain why they refuse to change something. Reading the work of these students quickly becomes a chore. You should have noted by now that it's all about being flexible for your reader, and being stubborn about anything you've written isn't going to work.

## 20.11 Rule 11: Read it yourself

Here's a rule that Scott didn't have. Perhaps it should have been inserted higher up the list. It would help you to have your work read through by one of your colleagues before you give it to your advisor. But above all, you must be prepared to read **all of your work** yourself, and be prepared to work on it when you read. Edit and improve it on every read. Not just the first time you write it, but for **every** version. Please don't ever expect your advisor to read something through when you can't be bothered to read it yourself.

Being concise is a great way to write well, and the abstract is a case in point. Please don't forget to read Scott's ten rules as written by Scott (Hotaling, 2020).

# 21

## Writing a PhD If English Is Not Your First Language

Reading this book, you will learn of some biases in biological sciences and how you need to be aware of them in order to properly interpret your work. Be aware also that biological sciences is biased in where studies take place, and who publishes them (Culumber et al., 2019; Nuñez and Amano, 2021). What we really need then are more studies from less studied parts of the world, and we need these to be researched and published by people from those countries. Why then do we insist that science be written in English?

English may be one of the most difficult languages to learn write as a foreign language because rules are so inconsistent, and the vocabulary is so large. At 171,476 words, English has an enormous vocabulary, followed closely by Russian (150,000), Spanish (93,000) and Chinese (85,568: although you should note that counting words isn't very accurate and as for counting dictionary entries, Korean wins). I cannot defend the use of English as the language of science, but I can assert that having a universal language for communicating science is massively helpful, and makes a large swathe of the literature available to those scientists who are prepared to become proficient in a second language. However, none of this really helps you to prepare writing in English as a second language.

One of my 'pet hates' is that many non-English speaking editors seem to insist that you have your work checked by a 'native English speaker' (see Figure 21.1). In my experience, having advice from a 'native English speaker' is no guarantee to getting a well-written chapter or manuscript. Most of the world's scientists did not grow up speaking English. Yet, rightly or wrongly, English is the language in which science is currently written. So, if English isn't your mother tongue, should you expect to receive help during peer review?

DOI: 10.1201/9781003212560-23

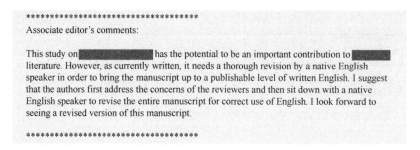

**FIGURE 21.1: The message from the editor: "Your work should be checked by a native English speaker."** Should you expect to receive help with your English when you submit a manuscript?

## 21.1   Message from an editor

Perhaps I should start from the outset by stating that English is my mother tongue, and that I have spent many hours correcting the language of colleagues for whom it wasn't. However, most of these hours were spent when I was a postgraduate student or postdoc. I no longer think that correcting English is my role either as an editor, reviewer or as an advisor.

As I've stated above, I did not study English, and would be the first to admit that my grasp of grammar and syntax is by no means perfect (those of you with better English will have picked this up by now!). I have read enough correct English to know when something is incorrect, but that doesn't mean that I know how to correct it. I have spent many long hours trying to decode what others have written, and in some cases, this has involved me rewriting entire manuscripts. I still do this as a co-author, although I do remember asking one colleague to please send any further drafts in their native Spanish as it was easier to translate than to re-write it.

The time component is at the crux of my reasoning why, as an editor, reviewer and advisor, I will not provide an English language service. It is time consuming. We all have our own voice, and correcting while maintaining other people's voices is a painstaking task. There's an entire profession that specialises in this (think translator). These days there are also services available from publishers to non-English authors to help them correct their English.

In addition, there turns out to be research suggesting why peer review is not the best way to improve writing (Shashok, 2008). Peer review works much better at screening technical content than it does at improving the communication of that content. And this agrees with the opinions of some non-English speakers who ask for editors and reviewers to concentrate on the science, a sentiment with which I agree.

No matter how you might hope that it isn't true, a poorly written manuscript will not get a good review. If your reviewer is struggling to understand what you have written, this becomes the overall impression that they will likely pass on to the handling editor. I try to separate my difficulties with English from my review of the science. But this isn't always possible. Frequently, a poorly written manuscript will mean that I won't be able to understand why the research was undertaken, what was done or what it means. This is bound to impact the review negatively.

The more I struggle to read, the more negative the review becomes. I see this as inevitable. What shocks me is that some senior researchers consider it to be their right to submit poorly written manuscripts and have reviewers or editors correct them (if you don't believe me, see Statzner and Resh, 2010). Worse, I've received manuscripts that are co-authored by people I know are native English speakers but are full of glaring mistakes that appear never to have been checked. For me, this violates the terms that all authors have approved the final manuscript.

But I've also got to admit that it's not easy. There are other places where you might be better to go to get advice on writing a paper or thesis as a second (or even more!) language (see Firth, 2017). I'm full of admiration for those of you that have to do this.

---

## 21.2   Must a 'native English speaker' check your work?

Being a native English speaker, I am not above having people complain about my English. I do get it wrong, and my English can often be improved. I'm always happy to receive help, and see it as a sign of how I can improve the clarity of a manuscript. However, I hope that my manuscripts are never so poorly written that a reviewer or editor cannot make sense of them. So it is a matter of degrees.

There are many people who are not native English speakers who write far better than I do. My 'pet hate' is that reviewers and editors insist that a manuscript must be corrected by a 'native English speaker'. I've seen so many very poorly written essays, theses and manuscripts written by native English speakers that I know that having one correct your manuscript is unlikely to be of much help, unless they themselves are good writers.

Being a 'native English speaker' doesn't automatically qualify you to write well, edit well or do any of the things that non-English speaking editors think that it does.

That non-English speaking editors often comment that a 'native English speaker' should read my text, simply underlines the problem that many editors themselves are incapable of knowing whether or not something is well written.

## 21.3 So if not a native English speaker, who should do it?

Anyone you know who writes good English and is willing to help you. One option is the service offered by **AuthorAID** (http://www.authoraid.info/en/). Using AuthorAID, you can find a long-term mentor who will help you with your English. You can read more about this approach here (Freeman and Robbins, 2006).

Failing that, I'm afraid that the best route will be to pay for help.

## 21.4 But surely my native tongue can be used for something?

Yes. It is essential that you use other languages to communicate your science with the wider community. Most of the world doesn't speak English, and so there is a real need for this to be done in your language. Many journals now allow abstracts to be submitted bilingually. Obviously, the second language should have some relevance to the study itself. Even if there is nothing in the instructions to authors about submitting the abstract in another language, I'd encourage you to do it. The more we scientists are exposed to the reality that other languages exist, the more it will rub off on those that need to understand. (Read more on this blog: Taşkin et al., 2020)

Kurt Vonnegut (1982): "No matter what your first language, you should treasure it all your life. If it happens not to be standard English, and if it shows itself when you write standard English, the result is usually delightful, like a very pretty girl with one eye that is green and one that is blue."

# 22

## Making Sure That You Don't Plagiarise

Plagiarism is when you copy somebody else's work. We're most familiar with plagiarism of writing and these days this is especially easy with the copy and paste function. Most people are not even aware that they have plagiarised because sometime in the past they copied and pasted the work of somebody else into a document and this later became incorporated into their text without them being aware of it.

Plagiarism is a problem because, essentially, you are taking somebody else's work without attributing it to them.

## 22.1 How to know if you have plagiarised

Today there are several pieces of software that are used to scan text that's written and available on the internet to discover plagiarism. One such example is TurnItIn. One of the outputs of TurnItIn is to highlight text that matches other text already on the internet. As almost all journals publish on the internet, TurnItIn can accurately determine if text has been copied from another article or website. I usually set TurnItIn to determine plagiarism with five or more consecutive words.

Remarkably it is very difficult to come up with exactly the same words that someone else used to describe a phenomenon. Most people when they think about it feel that it wouldn't be that surprising if they came up with exactly the same words as somebody else.

When caught plagiarising most undergraduates claim that they simply read an article and then later happen to write the same words that were in the article. They categorically deny that they ever copied or pasted text from the article into their work.

**Try it. When you try it you will learn what plagiarism is all about.**

Read an article, and then try to write text that is exactly the same as that in the article without looking back at the article itself.

DOI: 10.1201/9781003212560-24

Unless you have an eidetic memory you will fail at this task.

**FIGURE 22.1: It's always tempting to copy and paste, but it is likely to lead to plagiarism.** Copy (Ctrl + c) and paste (Ctrl + v) have become so easy that it is tempting to pick up portions of appropriate text directly from papers and then slot them into our own work (Figure 22.1). However, this is plagiarism and can easily be found by using software like TurnItIn. Most institutions will require checks for plagiarism on your thesis after submission, with dire consequences if your text fails.

This is not to say that no five or more words can ever be the same as someone else's. There are situations in which this happens. Think of addresses, quotes, certain laboratory equipment or protocols, and certainly references at the end of your paper. So there are many times when TurnItIn will come back with matching text. This is not what we're looking for in plagiarism.

## 22.2   What to do if plagiarism is detected in your work

It's remarkably easy to remove plagiarism from your work.

**Here's what you do:** Read the sentence that has been plagiarised several times to yourself. Now without looking at that sentence, write another sentence that has the same meaning. Because it's very hard to replicate somebody else's words without copying them, what you should find is that you've written a fresh unplagiarised sentence. This can now be added to your text, changed as appropriate to fit your existing text. And that should be the end of your plagiarism worries.

**Here's what not to do:** Take the sentence swap out some of the words for synonyms and pass it off as your own. The sentence will still have the same structure that you copied and, essentially, this is still somebody else's work. Moreover, TurnItIn will still recognise this as plagiarism.

## 22.3   How can you make sure that you never plagiarise?

Quite simply, if you never cut and paste, you won't plagiarise. It's that easy. I understand why people copy and paste as a way to get started, or because someone else has written something so well, it's hard to believe that you could ever write it any better. But actually, you can write it just as well, and writing it in your own words is worth so much more.

Don't forget that the penalty for plagiarism in your thesis might well be that you fail.

I still find it gratifying when I submit a student manuscript to TurnItIn and see that it is completely free of any plagiarism. There are so many words in English and so many different ways of putting them that you really can have your own writing style. Your writing style will be as unique to you as a fingerprint, and it will be entirely free of plagiarism. It's something you can celebrate.

Want to read more about plagiarism to understand what it is: see Louw (2017).

## 22.4   What is autoplagiarism?

I would maintain that there is not really any grey area in plagiarism. So what are people talking about when they refer to text recycling or autoplagiarism? The nominal grey zone within plagiarism comes when you copy your own text, this is sometimes referred to as self-plagiarism.

A new guideline from Hall et al., (2021) sets out the different types of grey plagiarism, or text recycling:

**Developmental recycling** is when you are reusing text that you have written for example between your proposal and something you intend for publication or in an ethics application that you also want to use in your thesis. All of this sort of developmental recycling is permitted and actually encouraged. I would further encourage you to use the opportunity of recycling this text to develop it and refine it further, condensing and improving where you can.

**Generative recycling** is where you take pieces of your already published text. For example, from the methods when it does not make sense to change the text or actually makes it more obscure to reword it in order to avoid plagiarism. In my experience this doesn't amount to more than a few sentences describing technical settings on equipment. However, this will depend strongly on your own subject area and may amount to larger chunks of text. I suggest that it is usually possible to reword most of the methods sections of papers. You should really only be generatively recycling material if you cannot avoid it: that is, when the text becomes more obscure by your attempts to reword it.

**Adaptive recycling** is where you are using your published text as the basis for a different type of content (e.g. a popular article online, a magazine, or op-ed). I think that this kind of text recycling is quite unnecessary because you almost certainly need to reword your text for a different audience. There maybe times such as figure legends where you need to reuse text that was already published. If you do find yourself in such a position then check with the copyright owner of the material that you are able to reuse the text that you want without legal issues.

**Duplicate recycling** (also known as 'manuscript recycling') is where large tracts of texts are essentially the same for the same message and audience. This is never likely to be sanctioned as it suggests that you are attempting to publish the same work twice. It will not be legal or ethical.

# 23

## Academic Phrasebank

It can be really tough to get started writing, especially if you are writing English as a foreign language. It could be that the academic phrasebank is just what you're looking for.

When reading papers, I often come across the same phrases time and again. They seem so well honed to the situation that I'm envious about how the writer dreamt them up. Stealing them wholesale would be plagiarism, but then they don't regularly fit the exact situation that I'm looking for.

Imagine cutting and pasting these phrases into a file, or even better a database that's ready for any eventuality. Banking them over time so that they are always available.

Then imagine that you did this and made it available to everyone...

Well, John Morley has already done it, and he's made it available to everyone. The academic phrasebank is a fantastic resource:

**the academic phrasebank**[1]

## 23.1 How to find the phrase that you need

- Look across the top for different sections of text that you are currently writing.
- Then look down the left-hand side to see the type of language function that you need. Click on something that looks applicable and watch the database open up.

---

[1] `http://www.phrasebank.manchester.ac.uk/`

DOI: 10.1201/9781003212560-25

## 23.2   Getting started

It strikes me that this is going to be a fantastic resource for those of you who are not writing in English as your home language. And for those of you who just need a few words to get started. It'll also be useful when you are wanting to wordsmith your introduction or abstract.

Is it plagiarism? The preceding chapter is all about not copying and pasting, am I now telling you that it's ok to copy and paste? Most of these phrases are less than five words or you will find that they are non-consecutive parts of a sentence. I don't think that most of these would constitute plagiarism if you pasted them into your document. However, I would suggest that you spend some time refining what you've written and that when you do that, some of these phrases may then come out.

# 24

## Why Critical Reading Is Crucial for Improving Your Writing

It is hard to emphasise exactly how important your reading will be, when it comes to your writing. If you are sitting with a blank page in front of you and feel that you have nothing to draw on, then think again. All of the reading that you have done to get to this point has already helped you more than you think. As I've pointed out elsewhere, your chapter or manuscript is likely to resemble the formula that many others already follow, and standing on the shoulders of those giants will help you again. But it's possible that your reading isn't helpful if you aren't being sufficiently 'critical' when you read: and by critical, I am talking about the methodical approach that you take to reading that improves your comprehension of a paper (Figure 24.1).

**FIGURE 24.1: Critical reading is seen as being made up of five important aspects.** This diagram shows how each of the five aspects feeds into the next.

DOI: 10.1201/9781003212560-26

## 24.1 Making notes on good and bad styles

When you read, it's worth making all sorts of comments about what you're reading. It's tempting to think that the only aspects of a paper that you should take notes on are the science (questions, approach and findings), but this is only part of your mandate as a critical reader. Of course, you should be taking notes on the science, but you should also be using it as an opportunity to learn tips and tricks of writing. Once you are actively writing, and I'd suggest that this should have been right from the very beginning of your studies, you should be actively thinking about how the authors are writing. Are they writing well? Are parts written poorly? Then the critical part is to cogitate, albeit briefly: what works, what doesn't and why.

Some papers will only offer the science that they contain. Others will stand out because you'll realise that you enjoy reading them. As a critical reader, your task is to decide why you enjoy reading a paper:

- Is it the style?
- Is it the way in which they formulate the logic of their argument?
- Could it be that they've broken the standard formula?
- Is it just the title, or the first sentence that you enjoy?

Once you've decided what it is, make these notes on the paper and file them on your referencing system (however you do it) under a style heading. I'd suggest that this be completely separate from your notes on the science. This is as simple as highlighting or making a quick note each time you see something that you like, or dislike (use different colours consistently). I used to underline text in pencil, and then make an illegible comment in the margin. It worked for me, not because I could read the comment, but because I could remember writing the comment.

You can also learn from badly written papers. Again, ask yourself exactly what it is about the writing style that doesn't work for you? If you don't know then comment in your database and make a group of papers that don't read well. Do they have anything in common?

It won't be every paper you read that you make notes about the style. It will probably only be a minority of papers in your area. However, this critical reading where you make notes on the ones that do stand out will help you greatly when it comes to your own writing.

## 24.2   Some examples

- Instead of using a taxonomic list (I don't like lists) and long citation string to describe lots of different studies that had previously made the same point, a co-author (in a recent review study) organised them all into the same categories we were presenting in a review. The result was that the reader was reminded of our categories at the same time as seeing how well others had already covered these same topics.

- Being able to sum up a paragraph with a clincher sentence that also has a segway into the next paragraphs is a great skill to have

- Some authors are able to generate a theme that they develop over several paragraphs or pages

If you've spotted what it is that you like, write it down and commit it to memory. The very fact that you start to see things in other people's writing that you like and admire means that you've managed to start reading critically. Keep this up, and it will help your writing no end as you open up your mind to critical reading.

You don't have to restrict yourself in what you can learn while reading other people's papers. Think about the stages in writing a paragraph. Look at the paragraphs you are reading and see if you can spot the subject and summary sentences. Maybe they are absent from the entire paper, or maybe they are present in every paragraph. Does it help enhance the readability of the paper? What about the times when authors break the rules? Can you see in your reading that the majority of authors follow the rules? Do they follow the formula? What is the power in shaking it all up? If it works, try and analyse why it works. Then you will be reading critically.

## 24.3   Make handwritten notes

When I started reading papers, I made photocopies (1 sided was as sophisticated as it got) from volumes of bound journal issues, and made notes in pencil on the back and front of those pages as I read. I understand that most of you will be working on pdfs which offer possibilities to add notes in text boxes that go alongside. But there still may be more advantages to writing notes using your own handwriting. A study by Mueller and Oppenheimer (2014) showed that students who typed retained less of the content from a lecture than those that made notes by hand. They showed that this was because those who used a keyboard tended to take notes that were more shallow, while those that wrote their notes had deeper thinking. My take on this is that the brain is making extra connections when you are taking notes by hand. Those extra connections will benefit you when it comes to trying to recall the study.

Another advantage that making notes by hand brings you is that you can deliberately associate the location in which you read, and practice your hand-written notes, with the place where you want to eventually write and recall what you've learned (see Bouton, 1993). Thus you can manipulate this facet of human memory to do all your learning, for example, at your desk in the department, or in the university library. Eccentrics may want to link it to reading in the bath, or another place of great comfort. But be careful not to make your 'place' too specific. You could move house or have someone else allocated to your desk, and that wouldn't work out well. However you decide to make your notes, make sure that it works for you both in the present, and for the later stages of your PhD studies.

Similarly, if you make notes on style, you are likely to remember them in the same way that you remember the paper and its contents. An alternative to writing everything onto a printed copy is to make notes in a notebook. That way you get the advantage of building your memory with your own writing, and you can easily find everything in the same place when you forget where you read the fascinating insight. Alternatively, you may want to print the most important papers in your area, and make notes in pencil all over them.

It's important to emphasise that you should not copy and paste parts of someone's paper that you like. It's easy to do, but any copying could be your undoing if you later forget and paste your 'notes' into your manuscript. It's so easy to plagiarise like this that the only way to really avoid it is by being very strict with yourself. Making notes by hand will force you to look away from the written text while you write, and you'll be unlikely to inadvertently copy anything. The advantages of writing notes by hand then make it greatly advantageous to restricting yourself to your computer (tablet or cell!).

## 24.4   Recognising good or bad in what you read

It's going to be hard to read and think at the same time. A little like walking and chewing gum, it comes with experience. And if you do fall over, stop chewing for a while and wait until you've got into your stride before having another go. Thinking is part of the reading process, and while I agree that it might be hard to think on multiple levels while you read, it is possible. For example, consider what other worthy things you've thought about while reading this book. If you've not stopped to check your phone, or think about eating or drinking something, well done. It might be worth creating an atmosphere with less extraneous stimulants while you read. If you only have your reading material and notebook, you'll be able to concentrate all of your thinking on them instead of getting distracted.

If you can't decide what it is that you enjoy from one paper, take it further. Take a look at the first author (likely to be the person responsible for the writing style). Have they written other papers, and do these have those same qualities? Or maybe it's the lab that they came from? Try to find the lab's pages on the internet and see if there are other inspiring papers there.

There is also a possibility that you are struggling to get insight with your reading because English isn't your first language. The subtleties of some of the usage of English may be passing you by. Because of the large number of words, it is possible to write something in English in many different ways. The overall aim is to convey the largest amount of information possible in the smallest number of words, while enhancing the context, meaning and readability. It's not easy, but the only way you'll ever get good is by starting, and reading and writing as much as possible.

If you are still struggling, a really good way forward is to start a journal club with your colleagues. It is probable that you aren't the only one who is having difficulties. There may already be a journal club in your institution which may (or may not) be a good place to go. What you do need is a safe space in which you aren't afraid to say when you don't understand. Think of it as a book club for scientific papers. After all, it may be that no one understands and the paper is badly written. Alternatively, it can be that someone is able to help you quickly and easily. They can then give you extra insights into how and why they think the author is crafting their paper. This will then help you with your reading as well as your writing.

If you are looking for good fodder for your journal club, consider reviewing a preprint and publishing your review with a DOI (Teixeira da Silva et al., 2017). This will make the whole exercise a lot more worthwhile and produce a net output for your club.

# 25

## What Is Needed for Your Research Proposal?

In my institution, you need to have your PhD or MSc approved by a group of 3 independent staff members (i.e. not including advisors) in your department. The department requires that your written proposal is publicly available to everyone in the department at least seven days prior to the oral presentation of your proposal. This is not unique to my university, many institutions require that you formally propose the work that you plan to do in a PhD. They may all have very different requirements. It is well worth checking out what your institution requires and by when. Talk to your advisor about this.

Your proposal is a collection of documents, where each document is a chapter of the thesis. Ideally, these should be bounded by an introduction, which gives the bigger picture of how the thesis is placed, and a timeline at the end that demonstrates how you can complete the thesis within the two or three years for your MSc or PhD, respectively.

### 25.1 Structure

Your final thesis will likely be a set of papers (data chapters sandwiched between an introduction and discussion), and so I suggest that you prepare the proposal in the same way. That is, you will normally have two (MSc) or five (PhD) data chapters. In the proposal, each chapter will consist of three parts:

### 25.2 Proposal introduction

I'm not going to go into what's inside the introduction here, as it's already been done in other chapters on formulaic writing (see later in this part) and specifically about the introduction (see Part III). Although you may feel that there's no point putting a lot of effort into this introduction it serves as a literature review for your upcoming PhD study, and is well worth the time and effort.

## 25.3  Methods and materials

What's in the Methods and Materials section is also covered in Part III. Essentially you need to:

- Introduce important aspects of your study organism and/or study site
- Describe exactly how you'll set up your experiment and/or collect your samples
- You must explicitly state how each of the variables (introduced in the introduction) is collected
- (If necessary) Describe how you'll turn your collected data into the variables needed for your analyses
- Data analyses (how you'll test your hypotheses). This should include a description of the statistical or analytical techniques. It's useful to structure this section by each of the questions/hypotheses that you are posing

## 25.4  Hypothetical results

It's a very useful exercise to imagine how results will show that you have (or have not) accepted your hypothesis. You can't show results that would cover every eventuality, but give some typical scenarios that you think may happen if you accept and reject each hypothesis. This could be a bar chart, a map, a scatter-graph, etc.

## 25.5  Your proposal – what is it good for?

Once you're done with your proposal, you might feel that you've done an awful lot of work without having added anything towards achieving your PhD or MSc. However, it's actually a really useful document that you can use in a number of ways:

### 25.5.1 Copy and paste directly into your thesis

One PhD student estimated that she pasted 60–70% of her proposal text directly into her thesis. This included nearly all of the methods and materials section verbatim. Note that this **is not** plagiarism but developmental recycling.

### 25.5.2 Use it to raise money to do your studies.

Writing grant proposals is essentially the same as your thesis proposal, and so you can use this document (with some modifications to tailor it to the grant objectives) to raise money for your work.

### 25.5.3 Deposit the proposal to conform to Design and Analysis Transparency

This is the subject of another chapter (see Part III).

# 26

## Making a Presentation from Your Chapter, Paper or Proposal

In theory, it couldn't be easier to take your written work, be it a research proposal, paper or chapter, and turn it into a presentation. Many people find presenting ideas easier than writing about them, as writing is inherently difficult. On the other hand, standing up in front of a room of strangers, or worse those you know, is also a bewildering task. In this section, the aim is to break down the task of making a presentation into a number of smaller steps so that you can tackle each one in turn and make sure that you are ready.

1. **Give yourself plenty of time to prepare** Although your presentation might be weeks away, you are going to need ~4 weeks to get the presentation ready, run it past your advisor (and potentially other co-authors), and practice your talk. As usual, my advice is not to leave it until the last minute to prepare as you'll get out what you put in.

2. **Know who your audience is and what they expect** If you have no idea of who your audience will be and what they will be expecting, then attend some other talks in the same seminar series, or ask your advisor and/or other lab colleagues for input. The more cross-disciplinary your audience is, the more general you should make your talk. If you've still no idea, pitch it for a very broad audience at last year undergraduate level.

3. **Be sure of exactly how much time you will have to talk** Be aware that your time slot might include time for questions, and so find out exactly how much talking time you have. If there is a long time set aside for discussion, you might want to throw some extra slides in that can help facilitate answering expected questions.

4. **Make sure you know the format of the presentation** If you don't know, then you can't make any assumptions about what might be present in the room where you give your presentation. I turned up to give my talk with a presentation on a USB drive only to find that there was no computer, no projector and a room of people expecting me to entertain them for an hour. Even if you know that a projector is present, it's good to have some extra context:

- for example is the computer Mac, PC or Linux;
- should you bring your own laptop;
- is there a pointer and can it advance slides;
- if in doubt, then bring your own accessories.

5. **Know exactly where you are giving your talk and plan for plenty of travel time** Hopefully, you will be very familiar with the venue, and travelling to the venue will only involve a brief walk. I've made some rookie errors in the past, including deciding to cycle to the venue, getting caught in a cloudburst and arriving literally dripping wet (it was in the UK). The more comfortable you are with the whole setup, and the more organised you can be about getting yourself there without incidents, the better the talk will go for you.

Essentially, you have a story to tell, but that does not mean you are story telling. It means that your presentation will require you to talk continuously for your allotted period of time, and that the sentences must follow on from each other in a logical narrative; i.e. a story.

## 26.1 So where do you start?

### 26.1.1 Here are some simple rules to help guide you to build your presentation

- **One slide per minute:** However many minutes you have to present, that's your total number of slides. Don't be tempted to slip in more.
- **Keep the format clear:** There are lots of templates available to use, but you'd do best to keep your presentation very clean and simple.
- **Be careful with animations:** You can build your slide with animations (by adding images, words or graphics). But do not flash, bounce, rotate or roll. No animated little clipart characters. No goofy cartoons – they'll be too small for the audience to read. No sounds (unless you are talking about sounds). Your audience has seen it all before, and that's not what they've come for. They have come to hear about your research proposal.
- **Don't be a comedian:** Everyone appreciates that occasional light-hearted comment, but it is not stand-up. If you feel that you must make a joke, make only one and be ready to push on when no one reacts. Sarcasm simply won't be understood by the majority of your audience, so don't bother: unless you're a witless Brit who can't string three or more sentences together without [in which case you have my empathy].

## 26.1.2 A formula for presenting a proposal

1. You need a title slide (with your name, that of your advisor and institution)
2. Several slides of introduction

2a that put your study into the big picture
2b explain variables in the context of existing literature
2c explain the relevance of your study organisms
2d give the context of your own study

3. Your aims and hypotheses
4. Methods and Materials

4a Images of apparatus or diagrams of how apparatus are supposed to work. If you can't find anything, draw it simply yourself.
4b Your methods can be abbreviated but the methods are important for a proposal, especially the numbers of replicates and the variables that you plan to measure from them.
4c Analyses are important. Make sure that you understand how they work, otherwise, you won't be able to present them to others. Importantly, explain each of the variables that you introduced, and explained how to measure, and how they fit into the analyses. There shouldn't be anything new or unexpected that pops up here.

5. Expected results I like to see what the results might look like, even if you have to draw graphs with your own lines on it. Use arrows to show predictions under different assumptions.
6. Budget
7. Timeline

## 26.2 Know your story first and then make your presentation

Despite previous sections of this book cautioning you against storytelling, you really need to turn your study into a story in order to present it well. This means that you know what the message you want to reveal to the audience is. When you develop your story, be imaginative and creative. There are plenty of different styles to telling stories, and you should feel free to borrow styles that best fit your scenario. For example, you can reveal the punch line right at the beginning, and then work back through the presentation to show how you arrive there (but always try to keep something to reveal on the way there).

Props are also excellent to produce in talks, something that you can bring with you, and show to an audience, that reveals more about your study. You want your audience to understand your story, but you also want them to remember it. Anything that will help them remember is likely to be a good prop.

In order to keep your audience with you, you need to produce reminders throughout the presentation. This could be themed images or colours that correspond to particular concepts, questions or ideas. If you get confused and can't work out how to present all of the information, then the chances are that your audience won't be able to keep track either. Slim down your study to something that will be easier to put into a story. Just like a real story, some things can't be left out, like characters (i.e. variables) that you rely on to answer your questions. Don't take time to talk about anything off the prepared storyline, as you'll likely lose the audience or confuse them. Treat your main variables like characters that appear with some regularity and consist characteristics throughout the story.

Remember that you don't need to include every detail in your slides or your talk. If the audience can follow your story but wants to know the details at the end, then they can ask questions about the details. Indeed, you can tell your audience precisely this as a way of side-stepping chunks of (especially boring) methodologies.

The audience is there to listen to you tell your story, and your slides (if you decide to use them) should be props that help you tell your story. Talk and interact with the audience to tell your story. Point to the slides to help you illustrate the points that you can't otherwise make, or to underline important points in your story. Above all, you need to make sure that the audience is listening to you tell your story.

**Fewer words are better. Tell the story yourself using the props on the slide**

## 26.3   Slide layout

1.   Your aim is to have your audience listen to you, and only look at the slides when you indicate their relevance.
2.   You'd be better off having slides without words, then your audience will listen instead of trying to read. As long as they are reading, they aren't listening. Really try to limit the words you have on any single slide ($<30$). Don't have full sentences, but write just enough to remind you of what to say and so that your audience can follow when you are moving from point to point.

3. Use bullet-pointed lists if you have several points to make (Font 28 pt)

4. If you only have words on a slide, then add a picture that will help illustrate your point. This is especially useful to illustrate your organism. At the same time, don't have anything on a slide that has no meaning or relevance. Make sure that any illustration is large enough for your audience to see and understand what it is that you are trying to show.

5. Everything on your slide must be mentioned in your presentation, so remove anything that becomes irrelevant to your story when you practice.

6. Tables: you are unlikely to have large complex tables in a presentation as presenting raw data or small words in a table is a way to lose your audience. Make your point in another way.

7. Use citations (these can go in a smaller font: 20 pt). I like to cut out the title and authors of the paper from the pdf and show it on the slide.

8. If you can, have some banner that states where you are in your presentation (e.g. Methods, or '5 of 13'). It helps members of the audience who might have been daydreaming.

**Large font, symbols that are consistent, familiar themes and photos all make for a good slide**

## 26.4 Practice, practice, practice

1. It can't be said enough that you must practice your presentation. Do it in front of a mirror in your bathroom or in front of your friends. It's the best way of making sure you'll do a good job.

2. If you can't remember what you need to say, write flash cards with prompts. Include the text on your slide and expand. When you learn what's on the cards, relate it to what's on the slide so that you can look at the slides and get enough hints on what to say. Don't bring flashcards with you to your talk. Instead be confident enough that you know them front to back and back to front.

3. Practice with a pointer and slide advancer (or whatever you will use in the presentation). You should be pointing out to your audience what you have on your slides; use the pointer to do this.

4. Avoid taking anything with you that you might fiddle with.

# 27

## *Starting Out Transparent*

The scientific method requires us to pose a falsifiable hypothesis, design a controlled, rigorous and repeatable methodology, and report and interpret our results honestly. In theory, it's all pretty straightforward. I'd like to think that in a world where science is adequately funded, we'd have already reached a situation where transparency was endemic in the scientific system. However, science has not been so lucky, and where there are shortcuts, some people will take them in order to get ahead of others. One of the objectives of this book is to demystify what happens in biological sciences. In other words, the demystification in this book is only really needed because so much of what happens does so behind closed doors, and out of the gaze of less privileged scientists. As you read through this book, you will see countless examples of where things are not as they should be. I have always tried to insert solutions when I highlight problems, but these are admittedly piecemeal, rarely proven, and even in a best case scenario might be best thought of as temporary solutions. The logical antidote to all these shady scientific dealings is to turn to transparency.

We would do better moving to open and transparent science. The open science movement is gaining traction (Sullivan et al., 2019), and I hope that it will become the norm in the very near future. Those of you who are reading this now are likely to be students during an interim period prior to open science becoming mainstream. Therefore, you will be instrumental in adopting this process as soon as possible to ensure that science becomes a fairer place for scientists from all backgrounds, and with all results. The idea is that this will impact research all the way from planning to publishing in a set of Top Factors[1], which are likely to be increasingly adopted by funders (see here[2]).

In this book, I try consistently to encourage you toward transparent and open practices in science. Here my aim is to introduce this topic and provide an overview of the main areas in which you can currently make a difference by opening up your research. As the Centre for Open Science explains (COS: see here[3]), this will take a shift in the culture of the scientific community, which is why you need to understand and adopt a transparent approach from the

---

[1]https://topfactor.org/
[2]https://youtu.be/MLbgSGA581Q
[3]https://youtu.be/9YuNGB3vNOw

DOI: 10.1201/9781003212560-29

outset of your research career. There are much better and more comprehensive guides and online resources, so I encourage you to look through the literature cited here. Although I advocate hard for the COS, please be aware that this is not the only set of guidelines (see also recent moves by the Public Library of Science[4] and Peer Community In[5]).

---

## 27.1 Preregistration

You will spend a considerable amount of time at the beginning of your PhD studies planning what you want to do. In many institutions, this will take the form of a formal proposal. It will be agreed with your advisor, and may well pass the inspection of a committee. You will put a lot of effort into reading the literature in order to ask the best hypotheses possible. You will design your experiments with rigor and control, and potentially redesign them after your advisor and committee comment. The effort put into your proposal is totally disproportionate if it's never looked at again. Yet, this document is of historical importance because it says what you think before ever doing the experiments and collecting or analysing the data. Thus, it can in future prove that you were not conforming to confirmation bias, for example, by HARKing: rewriting your hypothesis after getting your results in order to get a significant result (Kerr, 1998; Forstmeier et al., 2017).

Confirmation bias is a problem in science because of the way that science is published. In brief, articles for journals with higher impact are regularly selected based on significant results. If you don't have that significance, they are unlikely to want to take your manuscript. For this reason, many scientists have sought to have significant results to report, and this is called confirmation bias. Confirmation bias is bad because it violates the assumption that we are answering the hypothesis that we started with, or it could cause scientists to manipulate their data until a significant result is achieved. For example, another form of confirmation bias is p hacking: repeating tests with different approaches in order to obtain a significant result. Rubin (2020) gives a good list and set of explanations. As there are a lot of known ways in which researchers have been thought to deliberately or accidentally report false-positive findings, one solution can be archiving your intentions, referred to as preregistration (Forstmeier et al., 2017; see Figure 27.1). If you register your proposal (or any research plan), you can present this historical document to a journal (probably

---

[4]https://plos.org/open-science/preregistration/
[5]https://peercommunityin.org/

4 to 5 years later) to show that you have done what you planned to do (see webinar on preregistration here[6]).

What have you got to lose? With a good proposal that you feel confident with, you have nothing to lose. By registering your proposal, you are most likely to gain, especially in the future if more journals require preregistered hypotheses and methods in order to submit to them. What this does mean is that you will need to do a good job of your proposal. This can be daunting when starting a new PhD, especially with respect to the analyses if you are not from a strong analytical background. Your proposal period should be time to make sure that your knowledge of analyses that you will do is sufficient. I would suggest that the best way to make sure that you are proficient is to obtain a working knowledge of how to handle a dataset such as the one you are going to generate in your experiment during your proposal period. If your lab already has similar datasets, borrow one. If not, generate some data that you can use.

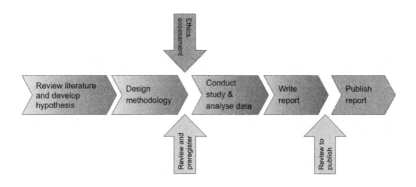

**FIGURE 27.1: Preregistration of your research proposal is a step towards transparency in research.** Preregistering your proposal formalises the first review process while publishing your manuscript formalises the second (after Scheel, 2020).

What if new analytical methods appear during the period, or you are forced to change your experimental protocol? Certainly, most analytical software will be updated over the course of your studies, and some might be superseded. None of this is a real problem to your preregistered content. Logging the software that you intend to use will demonstrate the approach and type of analyses that you intend to take. It is unlikely that you will ever be held to account for minor

---

[6]https://youtu.be/8QK2-udwoK8

changes to equipment or protocol as things do change along the way. It doesn't mean that you won't be able to report your results, or that your research won't be viable. But it does mean that you will need to be transparent about why it changed. For this reason, it is a good idea to document the changes to your proposed research plan and why they happened. It is surprisingly easy to forget! There are some great tools on different platforms for adding this information, together with a timestamp so that it's clear when it was done.

### 27.1.1   Does this mean that you can't perform any data exploration?

Although at first sight, preregistration appears to confine you to only testing ideas that you had before you started a study, actually, the opposite is true. When you preregister your proposal (or research plan) it places a time-stamp and, therefore, a historical trail on your ideas. If as you go along with your study, you gain new insight and ideas, you can register these with their own time-stamp. When it comes to analysing your data, you use a separate section labelled "exploratory analyses". You can then apply *post hoc* rules to these findings, which will not stop you from reporting them but gets around the awkwardness of HARKing. This serves to reinforce the distinction between prediction and postdiction.

### 27.1.2   Will preregistration of research eliminate the bias from science?

It is probably too early to tell (Rubin, 2020), but it is certainly a good place to start (Nosek et al., 2018). The more researchers know and subscribe to transparency in their research, the more it will shift the culture in science for the better (Forstmeier et al., 2017). However, there are more biases existing in science, and the system needs buy-in from stakeholders to promote equity by elimination of bias.

A new initiative from Peer Community In[7] is the possibility of submitting to their Registered Reports[8], which goes much further towards removing confirmation bias. The Registered Report (RR) is in effect the registration of a proposal (i.e. preregistration) with review. If the RR is approved by reviewers then the study is, in principle, given the green light for publication whether or not the hypothesis posed is accepted or rejected. I say 'in principle' because those same reviewers are shown the manuscript again once the results are in. They need to check that the methods proposed were followed, the analyses were conducted in the same way they were proposed, and that the conclusions

---

[7] https://peercommunityin.org/

[8] https://rr.peercommunityin.org/

are justified by the results. In terms of Figure 27.1, Peer Community In are offering to organise the two upward pointing review arrows. In addition, there are a bunch of journals that have already signed up to accept RRs that are signed off after completion (notable among these is PeerJ[9]). To me, this represents an important step in the right direction toward transparency, and the elimination of confirmation bias. What would be great to see is the number of conventional journals sign up with RRs based on the quality of the study design and execution, and the concomitant abandonment of Impact Factor as a driving force in publishing.

### 27.1.3    What platform should you use?

When choosing where to archive your proposal (or any of your data, analyses, etc.), there are lots of platforms to choose from: Bitbucket[10], Figshare[11], Github[12], OSF[13], Zenodo[14] and the list will undoubtedly grow. Making a decision about what you are going to use now may not require that you stick with this same platform for your entire career, but there are some things that you should consider:

**Here are some of the considerations that you should take on board**

- What do people in your lab and institution use? It'll be easier to use the same platform as your advisor and other lab members
- Some platforms require a subscription, check whether your institution is a member.
- Avoid using any platform that is tied to a publisher.
    - Although it's not possible to future proof this (as publishers have been shown to buy up anything they think will help them control the academic market), you can check how the platform is set up and opt for those that are non-profit organisations.
    - Other important aspects are "open source", free to use and open access.
- Can the platform function for more than one aspect of transparency?
- As the need for transparency in science grows, it will become important to have more aspects of projects archived. Does the platform that you've chosen cover all the stages from conception to publication?
- Is it easy to use?
    - Some of the platforms will be more intuitive to use, while others require a steep learning curve. Consider how friendly they may be to other collaborators, older advisors, etc.

---

[9] www.peerj.com
[10] https://bitbucket.org/product
[11] https://figshare.com/
[12] https://github.com/
[13] https://osf.io/
[14] https://zenodo.org/

- Are the archives easily compatible with other platforms? Working across platforms might be important for your project, especially if you start with collaborators that are already using different platforms.

The platform you choose should work for you (rather than having you work for it). If you are someone who loves to have everything ordered and organised, then you'll love seeing this all laid out. If you aren't, then these platforms are going to be a massive help to getting all of your plans sorted. Make updating your platform a habit. For example, you could make sure that notes taken during meetings with your advisor on different projects are logged onto the platform. This way, you both have a record of what decision was made when. Remember that you can choose what you make accessible to the outside world.

## 27.2   Transparency as you move forward

There are a whole lot more transparency criteria that you will need to be aware of later on in your PhD when it comes to publishing. Becoming familiar with the entire process now will be to your advantage, so I encourage you to read more about this topic. Sullivan et al. (2019) provide a nice overview about how to get started, but be sure to consult documentation at OSF[15].

---

[15]https://osf.io/

# 28

## Generating Funding for Your PhD Research

To be honest, writing a funding proposal is an art in itself and should be the subject of an entirely different book. What is written here is only really scratching the surface, so I suggest that if generating funding for your PhD research is going to be a large part of your work then you should look for an additional book.

## 28.1 Funding databases

Funding is one of the most important aspects of doing science, and something that a different emphasis is placed on in different parts of the world. In North America, in particular, some graduate students are expected to be able to demonstrate that they are able to raise their own funds, as this is expected of them in their jobs as academics. No matter where you are, your CV will be greatly improved by showing prospective employees (or labs) that you can generate your own funding. North America has a lot of opportunities to apply for funding for all sorts of reasons. There are less opportunities elsewhere, but the pool of people applying is also smaller. There are now so many opportunities, that several databases exist to help you find appropriate funding for your particular situation.

I am only going to address some of these portals, and to quickly consider some of the major reasons why you might want to apply for funding.

## 28.2 Reasons you might want to apply for funding

- To improve your CV. Even if it's only a small amount of funding that you apply for, it will make your CV look better. You might also want to search for prizes. Academic prizes carry both money and that feel good, look good feeling that CVs need.

DOI: 10.1201/9781003212560-30

- Your project is unfunded. If you are completely unfunded, or you have a bursary but no running expenses, then you are going to need money to do your work. The more work (particularly field work) that you want to do, the more money you'll need to raise.
- Your project is funded, but you want to do more. If you want to do a lot extra (like side projects or paying assistants to collect extra data) then all of this is possible if you can raise some more money to do it.
- You want to attend a conference or workshop. There are great opportunities for travelling around the world, but they cost money and international conferences are often very expensive.
- Publishing in Open Access journals. I find the idea of putting cash into the pockets of publishers abhorrent, but you may have little choice. There are some opportunities to get funding for OA publishing.
- It is clear that without making an application, you won't get any funding. But where do you start?

Databases. There are many databases, but here are three that are easy to use.

### 28.2.1   Mendeley: `https://www.mendeley.com/`

This is probably my favorite. It's easy to use and apply or remove filters. You can use the same login that you have for the referencing software and SCOPUS.

### 28.2.2   Open 4 Research: `https://www.open4research.eu`

For this you'll need an academic email address to register, but you'll get access to their research database. It is nice and logical to use, and you'll find local and international funding opportunities.

### 28.2.3   Research Professional: `https://www.researchprofessional.com/`

You'll need to create a login for this site, although you might get an in with your university address. This provides you access to what could well be the same database as those above. This is a slick database and relatively easy to use.

### 28.2.4   Opportunities in your department or societies

Probably the most likely place to get funding is where you already have an 'in' (where you are already known). This will include professional societies where you are a member, or your department or university. These places also have prizes, so it's good that you know what is on offer in order to benefit the most.

# 29

## Fear of Submitting Written Work

Many people feel nervous of handing in their written work, and feeling a little nervous is good. It shows that you care deeply about your work. However, when it gets so bad that you can't hand work in any more, then something has gone wrong. It is possible that you are suffering from **impostor syndrome**.

### 29.1 What is impostor syndrome?

Impostor syndrome is an experience you have when you believe that you are not as competent as you think others perceive you to be. It is not uncommon in many professions, and especially prevalent in academia (Clance and Imes, 1978). This is now widely recognised and there are lots of useful shared experiences out there to read (e.g. Dickerson, 2019).

- Are you fearful that your work won't be well received (Figure 29.1)?
- Is it really up to the standards that are required?
- Could you do better if you just spent some more time?

These are all really common thoughts, and they go along with impostor syndrome for a lot of early-career (and even older) academics (Woolston, 2016; Hutchins and Rainbolt, 2017). There are some important points to think about here:

1. Everybody has these ideas and you aren't alone
2. Handing in work and getting feedback is part of the learning experience
3. The fact that you care so much about your work and how it is perceived is a good thing. If you didn't care, then this would be a problem.
4. If you never hand anything in, you won't get your post-graduate degree

**FIGURE 29.1: It is common to worry about handing in your work.**
Many people feel conflicted about whether or not they should hand in their
work. Is this part of impostor syndrome?

## 29.2   Think of it this way

The process of writing is part of your learning process, and you aren't learning
alone. That's why you have an advisor. While you think that there may be a
great intellectual gap between you and your advisor, I can promise you that
there isn't. But your advisor is an experienced reader. Hopefully, there have
been many students who have benefited from learning with your advisor (talk
to them about their experience). They already work with a lot of students and
help those people to bring their writing to a level where it can be accepted by
an academic community.

Like other members of the academic community, your advisor also has experience of receiving critical feedback about their written work. Sometimes it is
painful and sometimes it feels personal. But getting feedback (or peer review)
is a fundamental part of the publishing process. Talk to your advisor and other
lab members about this process. Ask people how they overcame their fear of
submitting written work.

## 29.3 What can you do to help yourself overcome the fear of submitting written work?

1. **Give it to a friend or colleague to read**
2. **Remember that most of what you write should be understandable to the majority of people around you.** This means that you can give your written work to fellow postgraduates and postdocs and ask for feedback.
3. **Annotate your text with places where you are particularly unsure.** If there's a certain part that you are struggling with, annotate it (add comment) and point out that you are struggling with this particular section
4. **Ask for a meeting with your advisor.** You can sit down together with your advisor and discuss the points where you are uncertain before handing in the work. Or, after you've written it you can ask for a session when you are given verbal feedback
5. **Produce a checklist.** If you specifically worry that there may be errors in what you write (grammatical, spelling, plagiarism), then make a checklist that you can tick off prior to submitting. Once your check list is done, don't mess with the written work again (or you could add more errors), just hand it in!
6. **Set a deadline for yourself.** If you don't already have one, having a set of deadlines that you give yourself to give your writing to colleagues and hand in to your advisor. If you know that you aren't good with deadlines, share them with as many people as you can!
7. **Provide yourself with a reward for submitting your work.** Rewards are a really simple way of helping you to do things. They help simplify otherwise apparently complex fears.
8. **Tell your advisor about your fears.** There's nothing like honesty. Your advisor may be able to cut you some slack, or might sit down with you and look at how the two of you can overcome this difficulty. If you have more than one advisor, there may be one who you are more confident to read your work, and you can suggest that reading is done in series (instead of in parallel – actually, I'd advise this).
9. **Talk to your advisor about what really gets you the most upset.** In the case it's something your advisor did which made you anxious. If you can't do this face to face, then you could annotate it in a reply to their comment. The chance is that your advisor doesn't know how upsetting it is and you'll be helping them in their future interactions with students.
10. **Ask your advisor about their fear of handing in** and get them (and others in your lab) to share their stories (for example, at a lab

meeting). You might find that you have common ground to start sharing how these problems can be overcome.

At the end of the day, this is teamwork. Either you and your advisor are a team, or there are a bunch of advisors helping you. It is in everybody's interest that the job gets done. Getting a line of dialogue moving with your advisor is essential, even if you have to arrange a meeting about a different subject and then introduce the problem later in the meeting. There are deadlines that need to be kept. You need to find a way of getting through the fear, and if it's going to be a persistent problem, you'll need to work out what works best for you.

## 29.4   When is it good enough to submit to my advisor?

Once you've read and revised the draft text three times yourself, and you feel that you can't really improve it any more, hand it over. Personally, I need to wait a day or more between drafts when I work on something else, or do some reading. I find that it really helps to think about writing while not writing. Run through the arguments in your head when you are doing something else. That could be time in the gym, out running, cycling or swimming. Or talk through the logic of your argument with a friend or colleague while having coffee. You'll find that spending time thinking away from sitting in front of your word processor will be really valuable in promoting productivity once you sit down again.

If possible, hand it to an office mate or other friend first. Your reader shouldn't have to be an expert in your field but only needs to have a loose grasp on scientific writing to follow any paper. As a guide, any recent graduate of a biological sciences degree (BSc) should be able to understand. This makes it easy to find lots of potential readers for your manuscript before handing it in.

# 30

# Why Use a Formula to Structure Each Chapter or Paper?

Very few people can sit down at a desk and write an academic paper from beginning to end. For most of us, it takes a lot of hard work over many weeks or even months to write something that can be submitted for peer review. In the third part of this book, I approach the writing of a chapter or a paper for peer review in the same formulaic way, and will refer to papers or chapters interchangeably.

Sometimes it seems that every paper I read is just repeating the same formula over and over again. I even tend to forget that it's there until someone breaks the mould. Breaking the mould produces papers that really stand out, captivating the reader straight away. So why aren't we writing those manuscripts every time we write a paper? Moreover, why don't we teach students to write standout papers that will captivate everyone. Should we really be teaching formulaic writing?

Here I argue that not only is writing to a formula good practice, but it's the best way to learn scientific writing. When you look to see who is writing those knock-out articles (i.e. 'broken mould'), the answer is that they are really experienced researchers who have written hundreds of papers, and (unlike some of us) in doing so have learnt the essence of great writing. Such greatness is not innate and comes rarely to any researchers (even those who have written hundreds of papers). That's why I suggest to my students that they follow the formulas that are by now well recorded in many blogs and websites on scientific writing.

There is another reason why writing to a formula is recommended. Not only is it easier, but because it is so common and widespread, it also gives the reader a familiarity with reading your work. This means that editors are able to skim through and find exactly what makes your work worth publishing, and it means that those who are only semi-interested are more likely to find what they need to cite your work.

DOI: 10.1201/9781003212560-32

## 30.1   So what is the formula?

If we were in my office, at this point in the meeting I usually go to my white board and start drawing funnels (as a metaphor) and blocks. However, others have much better diagrams and here I've adapted an idea from Brian McGill's blog (McGill, 2016). But you can also find other useful material in other papers (Turbek et al., 2016; Plaxco, 2010; Tomaska, 2007). If you are of the Twitter persuasion, you get lots of useful tips and links from WriteThatPhD and Write4Research.

The diagram below (Figure 30.1) is supposed to show the movement of a liquid medium from one vessel to another. This is to make you think of your readers' understanding moving fluidly from one section to another. The introduction concentrates the ideas into your particular approach, while the discussion allows the reader to understand how your results fit into the larger body of knowledge.

### 30.1.1   My take on how to write a formula paper

But what goes into the top, and perhaps more importantly, what comes out of the bottom? I'd like for the discussion to end in a greater understanding of the system for the reader. But maybe if we could also inspire the reader to read more, that would be a great result.

My suggestion is that you now go and read five relevant papers in your field and annotate them to see how well they fit into the formula described (Figure 30.1). This will give you a practical idea of how the formula is put into practice.

### 30.1.2   Start with an outline

**So now that you've got the idea that writing to a formula is a good idea, what next?** Next, I suggest that you begin to plan out your formula as a bullet-point outline (Figure 30.2). Start by writing a general sentence for the subject of each paragraph. Then use sub-points to plan each sentence within the paragraph. Lastly, annotate these with citations that you want to use. Personally, I'd prefer to see the bare outline, and then a fleshed outline before you start the writing proper. This allows advisor and student to get on the same page to be sure that what is planned is thought out.

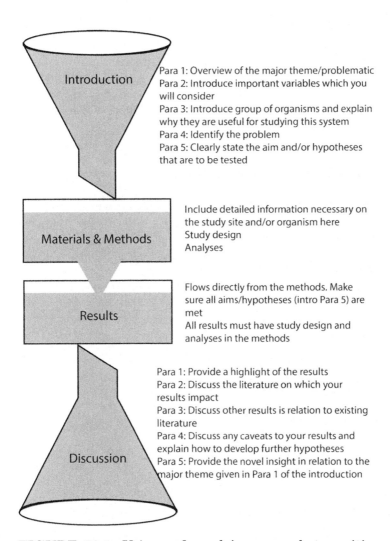

**FIGURE 30.1: Using a formulaic approach to writing your PhD chapter.** This diagram represents the formula of writing a paper or chapter that we will concentrate on in this book. Imagine the flow of the reader being like a liquid that passes from one section to the next.

In the above example, you may not know what EICAT and SEICAT are (they are scoring systems used for environmental – EICAT – or socio-economic – SEICAT – impacts of alien species), but I would hope that you can follow the broad argument of the introduction explaining why the study should be done.

### 30.1.3   Next flesh out the outline with citations

In the next stage, 'fleshing out', I ask that citations for statements above be added along with any examples. Figure 30.3 is an example of the citations added to paragraph 4 of Figure 30.2.

### 30.1.4   Lastly, it's time to write the text

And once you have the fleshed out outline, it's time to start writing your first draft. Remember that the best way to start writing is to do just that. It's unlikely that your first effort will be the one that you will finally submit. But start writing. Then read, go back and polish. And keep polishing until you achieve your goal.

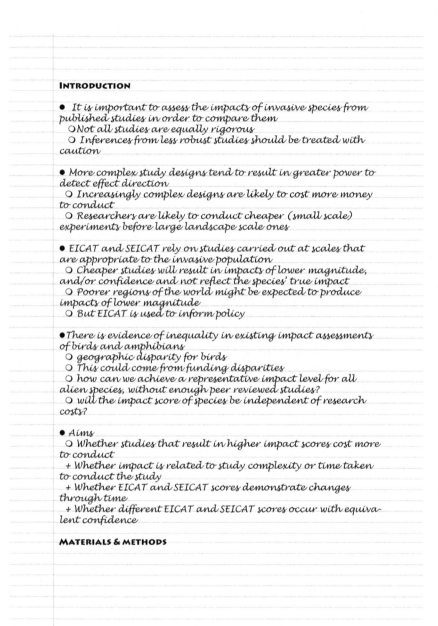

**FIGURE 30.2: A worked example of an introduction outline.** In this outline taken from Measey et al. (2020), paragraphs are indicated by the solid dots, and points to make inside the paragraphs by hollow dots. Each paragraph has an argument sketched out within it.

- *There is evidence of inequality in existing impact assessments of birds, amphibians and plants with lots of DD species* (30%; Evans et al. 2016; 38%; Kumschick et al. 2017; 79%; Rumlerová et al. 2016)

  ○ *Geographic disparity for birds* (Evans and Blackburn 2020)
  ○ *Examples of birds from Evans et al.* (2020)
  ○ *This could come from funding disparities*
  ○ *How can we achieve a representative impact level for all alien species, without enough peer reviewed studies?*
  ○ *Will the impact score of species be independent of research costs?*

**FIGURE 30.3: Fleshing out the outline of your introduction.** In this outline taken from Measey et al. (2020), citations and relevant figures have been inserted for each of the statements.

# 31

## Data Management

Almost everyone has had the experience of having written some really excellent work, and then had some terrible technology malfunction meaning that they lose everything. Some 'failure to backup' stories are truly horrific. However, they all represent a failure on the part of the person involved to properly backup their work.

At the time of writing, it is very easy to keep copies of documents in the Cloud, or even write straight into the cloud. Talk to your advisor about what options are available in your institution for backing up your work. Make a plan early on, before you start with your study, and then stick with that backup plan as you move through your PhD. A really good option is to start working in one of the Open Science platforms. For example, OSF allows you to create a project where you can continue to use lots of different Cloud-based resources, and provides an integrated home for them. OSF is also a great platform as it is free to use, and as long as individual files are less than 5 Gb, OSF will host them. This means that for most projects, raw data files can be stored on the Cloud at no cost to the project.

## 31.1 Should the cloud be the only copy of your data?

No. I suggest that you keep a hard drive with a copy of your data that you regularly update on a monthly basis, or after significant bouts of data collection such as field trips or experiments. A hard drive has the advantage that in the case of an internet outage, you always have access to your data.

Don't try to store data on a laptop or desktop computer. These are for working files only. Backup is a separate operation where you store files. While computers are good for computation, they are frail and much more likely to suffer from catastrophic failures. They are also objects that are prone to being stolen.

## 31.2   Backing up in the field

If you are going to be doing lots of field work, then you might want to make a specific plan about backing up data to the cloud during field time. It may seem expensive, but I'm sure that it won't be as expensive as losing the data. Alternatively, or in addition, you may want to take a portable storage option.

I still prefer field notes and data to be written into a book (preferably with waterproof paper such as 'rite in the rain'[1]). This allows you to make a photographic backup of the hard copy. Similarly, you can design datasheets that you use in the field with a pencil, and then photograph them to make a soft copy. I don't like entering data directly into a laptop or other device. For me, the advantage is that it is too easy to mistype a data point into a device, and then there is no record about what it should have been. These days there are multiple different ways to enter data into devices (including dictation), and so you should make your own plan about what you feel most comfortable with. If you do make a hard copy, you will then have to transcribe it later. If you make mistakes, you can go back to the hard copy.

## 31.3   Make a data management plan, and stick to it

However you decide to manage your data, make sure that you have a plan and that you stick to it. Don't become a victim of some unexpected event that will lose your data. These events will still happen, but at least your existing data will be safe. Creating a secure data management plan (Figure 31.1) is something that you should design and then talk through with your advisor. There is much more to data management than simply backing up your data. A great start is to read through the British Ecological Society's guide to Data Management[2]. See Part III for information about how to archive and distribute your data.

You probably won't be able to think about all the potential pitfalls that may happen during your PhD, but if there are some obvious ones, then do consider them and see whether you can mitigate the circumstances should they occur.

---

[1]https://www.riteintherain.com/
[2]https://www.britishecologicalsociety.org/wp-content/uploads/2019/06/BES-Guide-Data-Management-2019.pdf

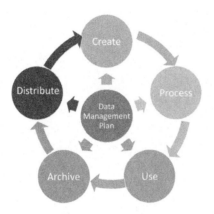

**FIGURE 31.1: The data management cycle.** It is possible to see your data as a cycle between creation and distribution. Your plan should aim to cover each of these steps (redrawn from the British Ecological Society's "A Guide to Data Management in Ecology and Evolution").

# Part III

# The Sections of Your Chapter

DOI: 10.1201/9781003212560-34

# 32

## Writing the Sections That Make up Your Data Chapter

Getting over the cognitive inertia of starting to write is top of the list for most of us. The approach provided here is to present you with a formula that breaks down each section into smaller more reasonably sized chunks. However, you still need to make a start. Procrastinating starting to write is normal. It seems a massive task, even if you've done it lots of times before. You want to write the best manuscript that you can, but you aren't feeling on top form today, so doesn't it seem a better idea to wait until tomorrow? It's not. You must start today and start now. Don't set the bar so high for your first draft. In fact, don't set any bar other than a personal goal of getting the first draft done.

For the first draft, type away confident in the knowledge that 90% of the first draft will get junked or changed so much that you won't recognise it. That's ok, and it's not a waste of your time. In fact, the first draft is invaluable in getting your ideas and perspective down on paper, allowing you to better organise and sift through your thoughts and bring clarity to your manuscript. Allow yourself a reward for finishing your first draft, something that you won't do until it's done.

You can make life easier for yourself by changing the order of what you write. Although I have placed the contents of this chapter as you'd expect to see them in a typical paper, this is not the order I would suggest that you write them in. The most important part of writing up your chapter or paper is to have your hypothesis or question clearly stated. Once you have this, it will help you to know the literature that is already there. Of course, you may already have your proposal to work with, in which case you can start by reviewing what you already have and deciding whether or not you need to tweak this, or if you need a fresh start.

The methods and materials are the easiest section to write first, especially if you've already done the practical work. This section will also help you focus on the analysis that you are doing for the results. Again you should review your proposal and see whether there were any significant changes to your planned methodology and note them down. This will not be an issue even if you have preregistered your proposal, you simply need to provide a rational explanation for why things changed.

Next, write the results. Remember that you are only looking to respond to your hypotheses, preferably using your (preregistered) proposal plans. Tables and figures are compiled along with the results. Now you have the materials and methods together with the results in full. I suggest that you make a list of all the relevant discussion points you want to make and then turn your attention to the introduction, and finally the discussion. For each, you should begin with an outline, before you flesh them out.

Lastly, write the abstract. Populate your article with references as you write. While you can leave the formatting of references until the end, it's dangerous to not put citations in as you write as you are likely to forget which paper is which by the end.

## 32.1 Title page

Most of the following information in this section on the title pertain to getting a manuscript ready for submission to a journal. For your thesis, I would recommend that you have a title page for each chapter that lists any collaborators and their contributions so that this is transparent for the examiners. It also helps break up the document and can be a good spot for a decorative illustration (if your institution permits).

Nearly all journals will require you to have a title page for your manuscript. This may or may not include authors. Check to see whether the journal that you are submitting to conducts double-blind review. If they do, they will require a title page without any indication of authors or institutions (they will also likely ask you to remove the acknowledgements section). This should be clearly specified in the instructions to authors of your chosen journal.

### 32.1.1 Names and addresses are important

Getting authors' names correct, their correct addresses (many people have more than one affiliation), can be tricky. Check that you have all of the information required for a title page before you submit. While you need the names of your co-authors to be displayed correctly, you do not need their titles (no Dr. or Prof., etc.).

Indicate clearly who is the corresponding author. Normally, if you have done all of the rest of the work, the corresponding author should be you. You need to learn how to start this role at some point, so it might as well be now. Having the correct name and address will be important to each author. Along with the author names and addresses you should record the ORCID of each author.

The ORCID is simply a unique identification code for individual researchers. ORCID is a non-profit organisation and there's nothing sinister in signing up. If they (or you) don't have one, then you should ask them to create one before the submission. It takes less than five minutes. Go to www.orcid.org

## 32.2   The title

The title of your paper, chapter or book is the first thing that any reader will read, and so should be well considered. If your reader cannot quickly understand your title, the chances are that they won't bother reading any further. Your title will be your selling point, and your aim is to use it to draw your readership in. Once you've managed to inveigle your readers to download your paper, your title is also their hook for remembering your paper in their database of thousands of others. Having their keyword in your title will help here, and as ever with writing your challenge is to think like your reader. The best titles are those that sum up the entire study in five to seven words. This is best done in a narrative that tells the story (see Part II) of your manuscript in its entirety. This may sound daunting, but you should get into the habit of summing up your story quickly (for friends, relatives and work colleagues). Then it's a question of refining this story into the short single sentence that makes up the title. While the narrative approach may not work for you, you do want the title to provide enough information so that the potential reader knows what they will find before they open it. Your title doesn't just have to work for you, it needs to work for a wide audience.

Some people are excellent at writing titles that contain puns of well-known phrases or sayings. These can be brilliant, working both to inform what's in the paper as well as providing some familiar input that helps retain them in memory. However, many fail to do either and are simply a waste of space. If you are tempted to use a pun as your title, make sure that it is widely appreciated, and not just among your co-authors and lab.

You don't have to come up with a killer title from day one. Most of my manuscripts have a working title that gets revised as I write, and is always open for change before submission. If you have great ideas for a title, do note them down. I find that the more options I have, the more likely I am to come up with something that works for everyone. It also helps to mix and match from a set of candidate titles. Once you have come up with something that looks good to you and your colleagues, test it by entering it into your database of choice (with the default being Google Scholar). Your first ten results should include a set of papers that you have likely cited in the upper area of your introduction. If you don't recognise any out of the top 10, it's time to look at another of your candidate titles.

### 32.2.1   Some title ideas to start you off

- Don't start by looking for the best title
  - Write a number of candidate titles and ask your co-authors to vote for their favorites.
- The shorter and catchier your title can be the better: 6 words ($\pm 1$) is an ideal.
  - Allow yourself a longer subtitle if needed, but don't go over 20 words total (some journals may limit your total to less).
  - Consider the (former) 120 character limit of a Tweet as an upper limit
- Do include your principal finding if possible
- Include as many keywords as you can

### 32.2.2   Things to avoid in your title

- Don't feel obliged to include taxonomic terms unless it is relevant or compulsory
  - Some (taxonomically minded) journals will insist on the species name followed by the taxonomic authority, and/or the family and order, in your title
- Avoid obscure or specialist words that won't be understood by your readership
  - There are times when keywords are necessarily specialised and your readership will expect this, but simpler words in your title will open up your readership which will otherwise remain narrow
- Don't simply define the scope of your work without including your content.
- Don't have your title as a question. Rather provide the answer!

## 32.3   Keywords

The keywords are a way for readers to find your content with searches. Typically, the advice is to use words that are not in the title or abstract. This is because many databases have combined searching facilities for title, abstract and keywords. I struggle to think of appropriate keywords, and so make a list of some of the big idea words and short phrases from the introduction. Then I tend to look at articles in the same genre, and see what keywords they have used. As with the title, I suggest that you enter your chosen keywords into your literature database of choice (see Part II) and see what comes back. You should see a group of papers that look wholly familiar and preferably those that are already cited in your manuscript. If not, it's time to look again.

# 33

## The Abstract

A good abstract is very important as, like a good title, it advertises the content of the paper and draws readers in. In a world where the quantity of scientific literature is increasing, it is more likely that someone will read your abstract but not your paper. Actually, it's far more likely that someone will read your title and use that to decide whether or not to look at the abstract. Moreover, when you submit your work to the journal, your editor may decide whether or not to immediately reject your manuscript based on the content of the abstract. Therefore, it had better be good!

### What is the abstract?

The abstract is a concise paragraph that sums up the major points of your manuscript so that the potential reader will be able to assess whether or not they want to read the entire paper. It is an abstract of the entire document.

## 33.1 So what would a good abstract contain?

A good abstract is a summary of the highlights of the paper. You can't hope to include all of the results, but you should include sample sizes, variables and relevant statistics when they respond to your major hypotheses. You must include the broader subject area that your study fits into, and show how your results are relevant to this.

None of this is easy, and you should not expect to write your abstract in a single sitting. It will likely require multiple iterations, and some intense word-smithing to make it as good as it can be. Abstracts almost always have a word limit, and that makes it challenging and means that you have to be concise see Part II.

Increasingly, you'll hear that a good abstract is citable. This means that it contains enough information that someone knows that it can be used to cite for a specific fact. Of course, these people should download and read the entire study, but there is no guarentee that they will.

DOI: 10.1201/9781003212560-36

## 33.2 Where do you start?

Just like planning your writing in general, I'd suggest starting your abstract with an outline. Use bullet points to make a list of things that you feel that you should include. Rearrange your list until you have all the introduction points at the start, results in the middle and discussion at the end.

There's no need for detailed methodology, but it is useful to know the approach. For example, 'we used a common garden experimental approach' or 'we sampled 85 animals from three invasive populations'. Although many abstracts are provided as a single paragraph, some are structured into the sections of the paper, or as numbered points. Using this formula is a good way to get disciplined about boiling your paper down into a small amount of concise words. It's worth keeping a copy of any abstract that you compose like this in case you decide to submit to a journal that requires it.

Starting with this framework will ensure that the abstract is well balanced.

At this point, I'd circulate it to your advisor to ask whether there are other key points that should be included. As a rule, it's easier to start with everything present, and only then cut the words down to something within the abstract word limit. If you wordsmith your abstract and then try and add a key point later, it'll never come out so well.

## 33.3 When do you write your abstract?

Although the abstract comes on page 1 of your manuscript, only try to write it once you've got to the end of the process of writing your manuscript so that you are aware of all the major points that you need to include. Maybe once you've submitted your second draft of the manuscript to your advisor (given that you are both happy with the content), you can formulate the abstract.

## 33.4 Do abstracts for conferences differ?

Yes. It's likely that an abstract for a talk will not be the same, unless you have already published the study: but even then it's probably worth re-writing it.

I'd suggest that your conference abstract be more descriptive and thought-provoking, such that it is aimed toward the specific audience of the conference. Pose questions that you will answer in your talk.

## 33.5 Where do people go wrong in writing the abstract?

- The most common mistake is getting the balance wrong. I often see an abstract that gets to the results, runs out of space and simply stops. It is important to have a statement about your interpretation of the results that relates back to your hypothesis.
- Another common mistake is to have a very simply worded abstract that conveys very little information. There is definitely going to be a lot to put in, so expect to write something too long with too much content, and then cut it down by discussing with your advisor what can be left out.
- You should never copy text from the main document into the abstract. You could start like this, but wordsmithing must change the words and their order substantially. Otherwise, when the first sentence of the abstract is the same as the first sentence of the manuscript, it will really turn off your readers.
- Too many statistics or their complete absence are also common mistakes. Adding numbers and statistics are important, especially when they take the place of more words.
- There is only time for one or two comments on the discussion, so pick out the most important. Look through your written discussion and rank the points that you make in terms of interest to others. Can you combine two or more points into a single sentence?
- The introduction of the abstract must frame the bigger question, and the discussion must show how your data responds to this.
- Don't be tempted to overreach with your claims in the abstract (or the main text). Include major caveats when they are relevant.
- Finish with something conclusive and strong, preferably how this study changes the understanding, and not a caveat or suggested further study.
- The abstract does not contain citations (or very rarely and only if unavoidable), so don't put any in!
  - If you have to have a citation in your abstract, then it must include sufficient information that it can be found without opening up the paper: *first author, abbreviated journal name, volume, first page.*

# 34

## The Introduction

## 34.1 Why do we need an introduction?

The introduction and discussion appear to be two common stumbling points for students writing chapters or manuscripts. First, what to put in and what to leave out. And second how to construct it. They are complementary, and if written well can be read without the Materials and Methods or Results, with the reader still gaining all of the major information of the manuscript.

The introduction is going to be the first part of your manuscript that anyone reads. Yes, they've already taken in the title and the abstract. These are almost like bait to draw readers in. The meat starts with the introduction. And it's not only important to get all the correct content in there, but it's also important not to give misleading information that might distract the reader.

When you write the introduction to a paper, you must not be tempted to stray away from what we are aiming to introduce. The aim of our introduction is to explain to the reader the hypothesis you are testing, and the approach you have taken to test it.

## 34.2 So what do we put into the introduction?

The hypothesis itself is made up of different parts, and each of these must be explained in the introduction. We also need to understand the approach that you've decided to take in your study (e.g. experimental, lab or field approach, observations or natural history). All of these decisions that you made were informed by the literature, as is your general understanding of the subject that you are studying. So making sure that you cite (read about citing) the relevant literature is a key ingredient of the introduction.

DOI: 10.1201/9781003212560-37

However, this is where I think many people get side tracked. Researchers love reading, and it's super easy to get sucked into all the amazing things that people have done in an ever expanding and increasingly interesting literature. We are often tempted to show exactly how well read we are. Or put in that fascinating tit-bit that we stumbled on by accident. However, you must keep focused on the goal, to introduce the hypothesis to the reader, and try not to allow yourself (and consequently your reader) to get distracted.

Consider this analogy: You can think of the introduction as being a highway to your hypothesis. As you drive the highway, it's fine to see signposts that lead to other places, but don't be tempted to turn off the highway to visit them. You do need to take your reader to your question as efficiently as you can. Point out the relevant sites on the way, but don't stop until you get where you're going.

## 34.3   And the construction...?

Previously, I've described the introduction like a funnel (Figure 34.1), where we channel the reader into our hypothesis by starting broad and ending up narrow (see the formula). This can draw criticism that journal articles are expected to fit a certain style, that this style is boring, and that what we really need is for people to write in a more exciting and varied way. Although I do not disagree with this, my objective here, is to try to demystify writing

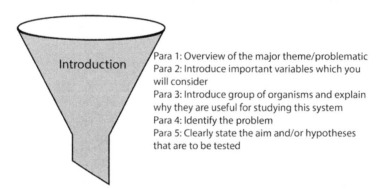

**FIGURE 34.1: The introduction funnel.** My suggestion is to keep to the funnel if you want to make life easier for yourself. Start by writing an outline of where you want your text to go. Then add in the references that are pertinent to each paragraph of the outline. Make sure that there aren't any paragraphs with a single citation repeated over and again; it's more likely that there is a lot more relevant information out there.

and enable you to get started. The easiest way to get started is, I believe, through a formula. But it is important to say that the funnel isn't the only way, and I've read some great papers where the first sentence of the introduction is the hypothesis. However you do it, the hypothesis is at the heart of your introduction because it is the reason for your work.

Note that you can shuffle your paragraphs in Table 34.1 to the point where it still makes sense to the reader. Don't be overly strict or dogmatic with this (or any) advice. For example, there may be more than five paragraphs, but use the framework to get started. Do what works for you in your situation. But beware of making any introduction too long.

**TABLE 34.1: Here is a guide to the contents of each paragraph of your introduction.** Although you should not feel constrained by this suggestion, it might help you get started when planning your bullet-pointed structure of the introduction before you start writing.

| Paragraph | Description of contents |
| --- | --- |
| Paragraph 1 | Overview of the major theme. |
| Paragraph 2 | Identify knowledge gaps in the field. |
| Paragraph 3 | Identify the problem and the gaps you intend to fill. Introduce the important variables that your hypothesis includes. It's not impossible to mention others, just don't get distracted. |
| Paragraph 4 | Introduce the approach that you are using, and the organism of choice. |
| Paragraph 5 | Clearly state the hypotheses that are to be tested. |

Now that you've fleshed out your outline with relevant citations, it'll be time to pass it by your advisor to check that you are moving in the right direction before you start writing. My suggestion is always to use your advisor to get the advice that you need – that's what they're there for!

Remember that while this section on writing the introduction is short, much of Part II concerns writing style that is applicable to your introduction. In particular, sections on composing your hypothesis, writing a paragraph, and an argument should be referred to when you are writing your introduction.

# 35

## The Materials and Methods

You should start writing this section in your proposal, and you may have to alter parts of it when turning it into your thesis later (if you change your methods). For most people, it's a relatively simple exercise, but here are some pointers to get you started.

The Materials and Methods section should be citation dense, especially if you have used standard methods that have been written up elsewhere. If there are different methods, you should also explain why you used one over the other, as reviewers or examiners may require this.

If you are introducing new methodologies, then you can expect to write a lot in this section about exactly how this was done, as well as the background to the reasons why the new approach to the methodology was taken.

Your aim is to produce a coherent methodology that anyone in the next 200 years (or more) can pick up and follow in order to replicate your study.

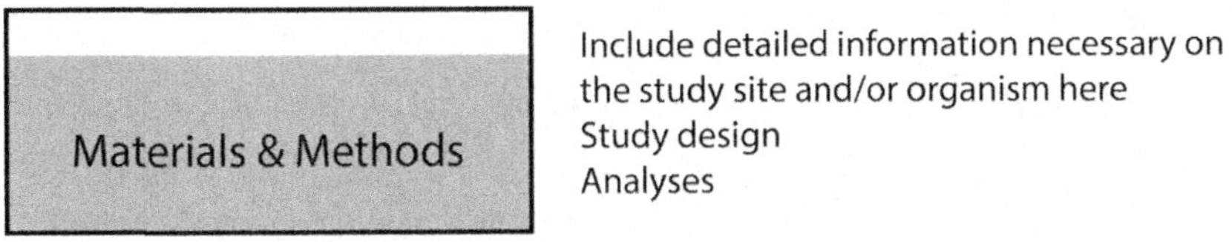

**Materials & Methods**

Include detailed information necessary on the study site and/or organism here
Study design
Analyses

**FIGURE 35.1: There's nothing to writing the methods except remaining methodical.** Your methods are likely to have some generic aspects to other study types in your field of biological sciences.

## 35.1 Sections or subheadings

As ever, I'm going to recommend inserting section headers into your Materials and Methods to help you break it up into bite sized chunks (see Figure 35.1). These are principally for the reader but will also make your job easier.

DOI: 10.1201/9781003212560-38

An important point here is to make sure that you keep your variables (what you are measuring to test your question) out of the sections like study organism and animal husbandry. Don't be tempted to scatter variables and how you measured them throughout the Materials and Methods section. Keep them clear and maintain a logical flow, using a time line if that makes sense.

## 35.2 Study organism or study system

Having already written the introduction, you will know that there isn't too much space to write a lot of information about your study organism there. However, often there is important background information that is needed on the species, which would be distracting or lose the flow of the introduction. By having a 'Study Organism' section at the start of your Materials and Methods, you can add in all relevant information in a paragraph or two.

For example, if you are working on African clawed frogs, *Xenopus laevis*, as a model organism you will want to use this space to describe why they are a good system for this particular investigation. Alternatively, your study could be detailing their invasion biology, in which case this would be the section where you can explain their distribution, that they are principally aquatic, or that they are invasive on four continents. You probably won't need all of these points, but just the aspects of this species' biology that is relevant to your study. You should make sure you know whether or not the journal requires the taxonomic authority. If you are not sure, include it now (see here for an explanation).

Similarly, if you are working on invasive zebrafish, *Danio rerio*, you can write in this section the known dates when they were introduced to Japan, and how we know where those animals came from. You can also give details of their natural range in South Asia.

If you are working on a community of animals, or it is the geographical region that is more important than any single species, you should explain the study system here. For example, if your study is about the rainforest, then you should provide some background information about this kind of vegetation here.

If relevant, you may need to provide a map with the location of samples that you used in this section. My preference is to try to provide a composite figure that will also include an image of your study organism (if relevant).

## 35.3 Animal husbandry / growing conditions / culturing techniques

If you have kept animals in the lab, grown plants or cultured bacteria in order to do your experiments, then you need to provide details of how you did this. For plants, this will include all relevant information on their source, propagation, growth medium and climatic regimes. For animals, you need to include their source, welfare, housing, feeding, light: dark cycles, temperature, etc. If you bred animals to produce your study life-history stage, all relevant information is needed here. All this information should be available from your ethics application. Some journals will want you to include the details of your ethics permission here.

## 35.4 Repetitive methodology

If you've used the same methodology for each part of your data collection (like DNA extraction, sequencing, etc.), you should have a section that explains all of this prior to any experimental manipulations done. I would suggest that this is detailed in the first relevant chapter of your thesis, and then referred to in later chapters. It's best to go to look to see what other people have reported on in order to know what level of detail to include. For example, 20 years ago, it would have been important to include detailed information of how you extracted DNA from tissues. Today, you can probably say that you used a standard extraction kit without saying more.

## 35.5 Do you need to include formulae?

Some of your methodologies will include calculations that rest on mathematical formula that can be written in mathematical notation. Certainly, there are plenty of examples in the biological sciences, especially in modelling, where these formulas are needed because they are developed within the methodology in order to produce the results. If your thesis involves this type of mathematical progression, then you will already know this is expected because all of the literature that you read will contain equations and formulas. However, if you are calculating (for example) the surface areas of a leaf through a set of measurements, do you need to provide the formula you used in your materials and methods?

I would move ahead on the premise that formulas that are well established in the literature need only be referenced, and not quoted. If you are building on any formula with new variables introduced by you in your study for the first time, then you will need to show the progression of these formulas. Remember that equations, formulas and theorems are all numbered in the text in a way that you can make reference to them later.

## 35.6   Experimental manipulations and data collection

If you have done more than one kind of experiment, it's probably a good idea to give each a separate subheading. Be consistent with these subheadings between the Materials and Methods and the results.

One very important point here is to fully explain the collection of all of the variables that are used in your analyses. Remember to use exactly the same names for your variables here, as you have in the introduction (and will use in your figures and results sections). It's very important that the reader gets this consistency across sections. Please also remember to include all units that you collect data in (there is a space between numbers and units!), and the accuracy of the measuring equipment used.

Your variables may be sufficiently complicated that you need to add extra subsections in order to accommodate and explain them properly. This is fine. It's also a good idea to use a table to explain a complicated sampling regime, or even a flow diagram. If you find yourself having trouble getting all of the information across in words, then go back to drawing a diagram that explains your data collection and ask around to see whether others find this an easier way to understand your setup. Diagrams and/or tables are there to replace words, so once you've decided which you want to include, you can delete the rest.

### 35.6.1   Equipment

Many journals require that you name the company that you bought equipment from as well as the town and country where it was made. This is to help others that might want to buy the same equipment (although it's a bit outdated in these days of multinational companies). You should try to get all of the model numbers of equipment as you do the work, so that you are not scrambling, later on, to find out what they were. It's really quick and easy to take a picture of this at the time.

If you have built or designed new equipment for your Materials and Methods, you can expect to make a diagram of this here, if it can't be adequately described in a paragraph.

## 35.7   Data analysis

This section is getting increasingly longer as people do ever more intricate statistical manipulations. As above, it's very important that you use the same names for the variables as you have introduced and collected them. In addition to explaining the specific tests that you performed, you should explain what roles the variables have as dependent, independent, or random variables in your model.

**The independent variable** has variation that does not depend on that of any other measure. These are usually measured by you during the experiment.

**The dependent variable** is usually the one that you are testing to see whether any of the independent variables explain it.

**Random variables** are subject to variations outside of your experimental control but that you want to make sure do not influence your interpretation of the dependent variable. Examples of random variables might be the order in which you did the experiment, or the position in your experimental setup).

You should be familiar with the terms above from when you formulated your hypothesis (see Part II for a refresher).

Any transformation of any of the data that you collected (e.g. log, ln, cosine) should be mentioned here, along with tests that you performed to ensure that they adhere to the conditions required for the statistical test that you conducted.

Give the name for each package in R that you use, together with its citation. Some journals will have particular software conventions that you need to respect and adhere to. Also, give a citation to R as well as the version that you used in your analysis. If you don't use the latest version of R, then your reviewers may want to know why, so always remember to update your version of R when you do your analysis.

## 35.8   Did you leave anything out?

By following your Materials and Methods, anyone else should be able to repeat your work. If there isn't enough detail for someone else to conduct the same experiment or survey, then you will need to add this information somewhere. For example, a list of sampling points or a database of sequences might be needed in Supplementary Information.

The Materials and Methods section can often get overly long, and is relatively easy to edit to make it much shorter. In journals where there is a word count limit, you may want to move some of the Materials and Methods into a supplementation information section. But in your thesis, try to keep the Materials and Methods section as detailed as you can.

# 36

# The Results

This is perhaps the easiest section of your chapter or manuscript to write. However, there are still many mistakes that are made. Let's start with the approach that you should take, and then go through some of the common mistakes.

## 36.1 The correct approach – responding to your hypotheses

The easiest way to write the results is simply to have the questions or hypotheses that you posed at the end of the introduction and have them as subheadings for your results section (see Figure 36.1). Once you have these subheadings, you simply assemble all of the results with respect to that hypothesis. You should already have all the tests that you used in the methods section.

The results section often includes a lot of statistical testing, resulting in lots of numbers inside brackets within the text. This is all expected and fine, but consider consolidating multiple iterations into a table (if appropriate). And if your tests are already in a table, then they should not be repeated in your text. Instead, simply make your statement asserting the results and add "(see Table 1)".

When describing the relationship between two variables, make sure that you state the direction of the effect. This means not simply stating that they have a significant relationship, but what that relationship is. This means stating that one increases with the other (a direct relationship) or one increases when the other decreases (an inverse relationship). Note that other types of variable relationship exist. Once you have described the result, either make reference to the Table or Figure where this can be seen, or describe the results of the test in brackets.

If you do have to insert the results of a statistical test into the text, include the test statistic, degrees of freedom (which incorporates N - the sample size)

DOI: 10.1201/9781003212560-39

and the value of P. Only if P is less than $\alpha$ ($\alpha$ is usually 0.05) can you use the word 'significant' when describing the effect.

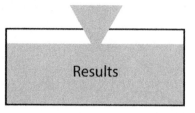

Flows directly from the methods. Make sure all aims/hypotheses (intro Para 5) are met
All results must have study design and analyses in the methods

**FIGURE 36.1: You may well find that your results section is the shortest section of your manuscript.** Only report what you need to in this section. Keep results that don't respond to your hypotheses for the Supplementary Information.

## 36.2 When to have a table or graph?

Most journals will allow you to have as many tables or graphs as you need to give your results. There are a few important things to consider. First, that data shouldn't be repeated. So don't have a table of results where the same results are also in a graph (or even two graphs that show the same data). Also, don't repeat data from a figure or table in your text. You need to decide how best to present your data. If you can't decide then discuss it with your advisor by presenting them with the alternatives. Good results should convey your findings clearly, and consulting with others should quickly help you decide which approach is effective.

See sections on tables and figures below.

## 36.3 Descriptive results

Some studies have a set of descriptive results that, by convention, are placed without a section header at the start of the results. These normally convey the size of the dataset that entered into the rest of the analyses. If the size of the dataset varies by design in the methods, then you don't need it here. But you can't always control the sample size, so it needs to be established before moving on to the rest of the results. In genetic studies, for example, this might include the number of samples that were collected and successfully sequenced, and thus entered into the rest of the analyses. Your text can then be broken into sections that correspond with your questions, unless you only have one, in which case further sections are not warranted.

## 36.4   Responding to hypotheses

While I encourage you to present data that corresponds to hypotheses here in the results section, you should not formally accept or reject your hypothesis in the results. This is done in the discussion, and allows you to present caveats and other important issues.

## 36.5   Common mistakes

- Having extra analyses in the results that you didn't include in your methods. If you really didn't anticipate conducting the extra tests, then these are actually *post hoc* (after the event) tests (not anticipated before the study) and you should think carefully about how to treat these. Otherwise, you'll need to re-jig your methods to make sure that all tests are included.
- Meaningless precision. This can happen throughout a thesis or manuscript (e.g. in the Materials and Methods), but it is most common in the results when too many significant figures are reported. For example, the output of your analysis might tell you that flight feathers in one population are 2.6484332 mm longer than those of another ($\pm$ 0.0654562 mm), but in your methods you only use callipers that have a precision of 0.01 mm. Reporting results beyond the precision whereby they can be measured becomes meaningless. Whenever you obtain the output of a statistical package, you need to make sure that you use your practical head to re-interpret these results and think about what they actually mean.
- Repeating numbers in the text that are already in tables or graphs. The text is likely to be packed with numbers already, so don't repeat any that are already presented elsewhere.
- Repeating text in different sections. As you've seen elsewhere in this book, repeating text simply isn't acceptable. It doesn't matter if you did the same thing many times, you can always write something in a different way. Having repeated text is the fastest way to bore the pants off your reader, and therefore must be avoided at all costs.

# 37

## The Discussion

The discussion should be the final section of your manuscript or chapter that you are going to write (given that the abstract will come last). This is because for the discussion you must already know all of the rest of the manuscript, starting with the hypothesis, which dictates what will go into all of the other sections. While writing the other sections, I often make notes under the heading 'discussion' so that it acts as an *aide memoire* to ideas that I've had during the study.

Before you start writing your discussion, make a plan and then discuss this plan with your advisor. I'd make this suggestion for all of the sections of your chapter or paper. It doesn't take that long to do, and it provides an opportunity for you and your advisor to talk about the results of your work and discuss them together. Such discussions should be eye opening for both of you, and they provide a great opportunity for you and your advisor to get excited about the work you've done, your results and what they mean. I get a lot of enjoyment during these discussions, especially when sharing the excitement of the results. Sharing thoughts before you start writing is important because by talking about it, you and your advisor are more likely to come to a consensus about the best way in which to interpret your results. Conversely, presenting them with the discussion finished might not be the best way of pitching your ideas.

There is a convention that you do not refer to the tables or figures in the discussion. This is because they have already been referred to in the results, and a reader should have already consulted them there. Occasionally, you may want to have a new conceptual diagram in the discussion, but this is rare. Most often, you do not refer to figures or tables and no new ones appear in the discussion.

In general, the first and last paragraphs of the discussion are key to the reader, but the discussion must also consider caveats and limitations in the experimental design and interpretation of your results, as well as providing a concise discussion of the results in the context of existing literature (see Figure 37.1). This is also your opportunity to suggest new hypotheses and how they could be tested

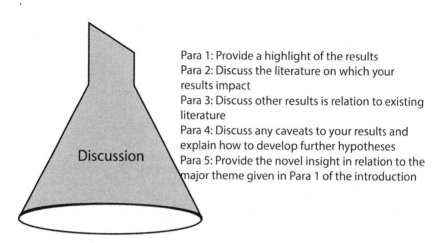

Para 1: Provide a highlight of the results
Para 2: Discuss the literature on which your results impact
Para 3: Discuss other results is relation to existing literature
Para 4: Discuss any caveats to your results and explain how to develop further hypotheses
Para 5: Provide the novel insight in relation to the major theme given in Para 1 of the introduction

Discussion

**FIGURE 37.1: Your discussion needs structure too.** Let's remember that if you are struggling to write, there is the potential to follow a formula, such as this one.

I'd like to repeat that the convention described above is a formula that is intended to help you getting started with writing. It is not the only way to write, and if you feel confident that another way is better then I encourage you to explore this with the support of your advisor. Be imaginative and don't feel that you are constrained or compelled to do anything in a formulaic way.

## 37.1   First paragraph of the discussion

Your discussion begins by you responding to your hypothesis, clearly stating whether or not you accept it, and putting this into the wider context of the study (i.e. paragraph one or two of the introduction with relevant literature). You can then follow these statements by emphasising what you consider to be the most important finding, and explain how it adds to existing knowledge. However, don't be tempted to over-interpret your results, or claim that they mean more than they do (see section on speculation below).

This first paragraph of the discussion doesn't have to be very long (three to four sentences), but you should make sure that you end by providing a link to the following paragraph or explaining how you will move the discussion on in sections.

## 37.2 To sub-section or not to sub-section the discussion?

My preference is to plan the discussion before you write it, just as you did for the introduction using a bullet-pointed list. This will provide you with logical sub-section headers for the discussion under which you can write the first draft. When your chapter has a simple aim that is easily communicated, I'd suggest deleting these sub-section headings before you finish. However, many studies are more complex and contain multiple experiments or evidential approaches. It is then sometimes wise to leave sub-sections in your discussion so that your reader can more easily follow the text. Where possible, these should be the same sub-sections that you have broken your methods and results into, especially where these relate to specific hypotheses or aims. Or it may be more appropriate to discuss the different approaches separately, specifically when the literature that you refer to falls into different groups.

If you are stuck and can't decide which way is most appropriate for your work, spend more time on fleshing the outline specifically to include the literature that you want to cite. Try it one way, and then the other, and you should quickly be able to tell which makes more sense. Of course, you should also ask your advisor for their opinion – that's what they are there for, after all.

When considering what sub-section to write first, go back to the order that you've presented the questions or approaches in the rest of your chapter or manuscript. Keeping the order consistent throughout is a really good way of helping your reader follow what you want to communicate. Shuffling the order in each section is almost guaranteed to get them lost and wishing that they hadn't started reading.

### Next you need to discuss!

The discussion is about explaining the meaning of your results to the reader. I often find that people write a lot of inappropriate information in the discussion. Remember that this section is not going to provide background information, and is unlikely to bring up new topics that need introducing. It may be that your results prompt you to introduce a new area of research that wasn't covered in your introduction, and this is fine. But for the main part, you should discuss your results in the context of existing literature. You can expect that the literature that you use in your discussion will only partially overlap with your introduction, with plenty of new citations. Similarly, it can be that discussing your results will mean that you end up with paragraphs that have no citations.

In order to provide a critical evaluation of your research findings, you really need to present a reasoned argument (Jenicek, 2006). In some respects, when writing the discussion you walk a tightrope between inferring the meaning of your results for a wider audience, and making claims that are too broad (Table 37.1). Overreaching in the discussion (and consequently abstract) is a common problem that examiners and reviewers regularly ask authors pull back on their claims.

When providing different sides of an argument, try to use your results to conclude that one side is supported more than the other. If your results don't help with this particular point, then it could be that you are trying to discuss something that isn't directly related to the work. This is a very common problem in discussions, and a good test is asking yourself how your results add to the point you are trying to discuss. If they don't leave them out and move on.

**TABLE 37.1: Some discussion dos and don'ts.** Balance your discussion so that you provide a reasoned argument that reflects what you found (adapted from Jenicek 2006).

| Do | Don't |
| --- | --- |
| Provide a summary of your major findings | Use statements that are too broad |
| Interpret your results with respect to mechanisms | Over-represent your results |
| Give problems with methods and techniques | Provide unrealistic solutions to deficiencies |
| Compare similarities and contrasts with other studies | Ignore conflicting literature |
| Provide practical implications of your results | Be unrealistic about what could be done |
| Give directions for future research | Simply state what you have done elsewhere in your thesis |
| Restrain your conclusions to evidence from your data | Make unsubstantiated claims that overreach your evidence |
| Provide insight into unresolved questions, controversies and biases | *Reductio ad absurdum* |
| Explain the study design's limitations and uncertainties | Inflate the importance of your findings |
| Highlight the strengths of your study | Forget to mention the limitations |
| Provide alternative interpretations and hypotheses | Go into tangential issues |
| Give unexpected results and anomalies | Avoid being critical of how your study could have produced unexpected results |
| Mention literature that contradicts your findings | Use your conclusions as a 'bully pulpit' |
| Provide arguments that are logical and coherent | Erect a 'straw man' or 'false dichotomy' |

| Do | Don't |
| --- | --- |
| Make conclusions based on your logic | Make emotional appeals to the reader |
| Make sure that you've discussed all results | Suddenly introduce new data or results not previously presented |
| Boast about your findings | Apologise or undermine your own work |
| Be concise | Waffle on |
| Take a look at target journal requirements and suggestions for a discussion | Feel that journal advice is not relevant to you |

## 37.3 Caveats and limitations

An important aspect of the discussion is to consider how the interpretation of your results may be incorrect. For example, if you have done an experiment, how well controlled was it, and how well could it be considered to scale up to real-world interactions? Could you have measured other variables? Almost every study will have caveats and limitations, and it is very important that you report them in a considered approach. You can also add ideas on how to improve your approach to eliminate important caveats that you have identified, even if this means suggesting something that adds evidence from a completely different field of biology.

My preference is not to provide all the caveats and limitations as a separate paragraph. Instead, mention them when you are discussing relevant aspects.

## 37.4 Should you speculate in the discussion?

Reviewers will often be unhappy with speculation in the discussion section. Speculation isn't that hard to spot, as it occurs when you make claims for which your results have no foundation. I think that it is healthy to have one or two statements that are speculative, but clearly label them as such. After all, after writing this paper, you are going to be one of the world experts in the topic, and thus your deeper understanding is often worth relating to the reader. However, I suggest that you speculate in combination with suggesting what work could be done in future. Remember, if you really feel that the point has to be made, you must clearly label it as speculative.

Perhaps an easier trap to fall into is over-interpretation. This is when you suggest that your results mean more than they do. It's an easy trap to fall into, especially after setting up the study in relation to key topics in the discipline (presented in paragraph one or two of the introduction). You will probably find it hard to see where you have over-interpreted, and this is something that having your work read by your advisor, or another colleague, will really help. You may then be asked to 'tone down' your claim, or to place it into the direct findings of your results.

Again, my preference is not to place all speculations or future hypotheses in the same paragraph of the discussion. These aspects should appear as the topics they relate to are discussed.

## 37.5   Don't beat up on others

Your results may show that other researchers were wrong with their interpretation or findings. Whatever you may think of them, never use your discussion to be disrespectful to other researchers or their work. This has been referred to as the "bully pulpit" or an *ad hominem* attack. As with all aspects of professional interactions, consider how you would like to be treated, and act accordingly. This is not to say that you shouldn't point out mistakes that were made before, but be sure not to be emotive or insulting.

Things brings to mind Freud's (1930) narcissism of minor differences, in which he pointed out that bitter fights could start between groups when they, in fact, agree on 90% on issues. So it is likely to be true that someone you might disagree with on one point, is likely to be allied with you on nearly all of their other views and opinions. Making and maintaining allegiances is more likely to get you in a position where you can constructively discuss any minor disagreements. Consider this before you are tempted to launch a negative sounding social media post.

Generally, such comments won't get through the peer review process, and remember that you might be insulting the examiner of your thesis (or the reviewer of your paper) – which is not likely to go down well!

## 37.6   Where next?

The 'where next' aspect of your discussion is important as it may provide the reader with ideas for their own work. Of course, these are questions that you may wish to pursue in your own career, or they may require corroborating evidence from other disciplines that you will never undertake. Either way, giving insight for continuing aspects of the research is an important component of the discussion. Providing new lines of research may also allow you to speculate about what you consider to be the most important angle of this topic now that you have presented your results. This should be justifiable and not gratuitous. There's no point in suggesting what other studies could be done just for the sake of it, or just because you happen to have already done them in your thesis. But if there are legitimate parts of your thesis that link to your discussion of your chapter (and this is quite likely), then you should definitely point to them.

## 37.7   Last paragraph

The last paragraph of the discussion is your take-home message. It's a summary paragraph that sets out what you aimed to achieve, and what the new state of understanding of the topic is now that your results are out. This should include the key literature that can now be reconsidered.

### 37.7.1   Never repeat text

Please remember that while this might sound similar to your first paragraph, it is not the same. This final paragraph should not replicate any text that appears elsewhere in your chapter or manuscript (not even the abstract). Never repeat or copy text generally, even within your own chapter (or between chapters). For the reader, it's very easy to spot and it gives the impression that you have nothing to say and are simply filling space. This is not the kind of impression that you want to give your reader, especially if they are examining your work!

As always, there are a number of other places to look for more advice to write your discussion, and I'd encourage you to read as widely as possible. For example, try Hess (2004), Jenicek (2006) and Şanlı (2013).

# 38

## The Acknowledgements

This is the part of your manuscript where you get to thank all of the people who have helped you along the way. It is easy to forget people and so it is worth considering making notes about who has helped you along the entire time with your PhD. Then all you have to do is remember where you wrote it down!

### 38.1 Thank your funders

You should thank any people or organisations that funded you or any component of your work. Check with them, as they may have specific wording that they want you to use. Remember that your institution may not have given you money but may have supported you in other ways, and you should probably thank them. There are registers of funders, and you should use these to see whether your funder is present and exactly how they should be acknowledged. One example is the Crossref Funder Registry[1]. Funders rely on being able to search databases for their name and find all of the work that they have funded. This means that you should use the particular wording that they supplied, especially if they specify this in their funding letter.

If your funders aren't in the database, you might suggest that they get added.

### 38.2 People and organisations to thank

- Try to remember all of those people who helped you doing fieldwork, or in the laboratory. Especially laboratory technicians (if they are not also authors) without whom a lot of the work that we do isn't possible.

---

[1] https://www.crossref.org/services/funder-registry/

DOI: 10.1201/9781003212560-41

- People that may have helped administer your funds or other office administration.
- Organisations that gave you permission to do your work. In some journals, the acknowledgements are the correct place to put your permit numbers along with the names of the organisations who provided them.
- Increasingly, permission from ethics committees is placed in the materials and methods but do check with the journal instructions to authors.
- People who supplied you with samples, images or other elements of your work or presentation of your manuscript.

Many journals have specific ways in which they want people's names to be written. You will need to follow these guidelines. For example, authors are often referred to by their initials in the acknowledgements. This is useful when some people in your paper need to thank particular funding bodies who didn't fund others. Thank anyone who gave you a photograph that you used in the paper.

## 38.3   Who not to thank (in your manuscript acknowledgements)

- You do not need to thank anyone who is an author. The exception is in your thesis acknowledgements, where you should mention them and their support.
- Personally, I tend not to thank people who were paid to do a specific job and only did that job. The acknowledgements is a section to name people who have gone out of their way to help you.
- Your family for getting you to grad school (unless they really helped you do the work). However, in your thesis acknowledgements, you should absolutely include your family.

If in doubt, it's probably a good idea to be as inclusive as you can in the acknowledgements.

You can refer to other acknowledgement sections in the same journal and see how they are formatted and written. Remember that if your journal is conducting double-blind review, they may ask for the acknowledgements (and the title page) to be removed from the main manuscript and uploaded separately.

# 39

## The References

Having the references at the end of your chapter or publication is actually very important because in-text citations are often not sufficient to determine exactly which paper you are referring to. Once you become very familiar with your own specific field you will find that you do glance at the references to make sure that the authors are citing the papers you think they are. It's also a great place to learn about literature that you don't know about already, and for this, you need the full citation in order to look up the paper.

The right time to make sure that references are in the document is when you are writing the document. Do not leave this until the end. It has happened to me many times that I have written an in-text citation, did not paste the citation at the end of the document, and by the time I had finished writing, had forgotten what I had read and where I had read it. This then leads to a lot more time wasting trying to find that or another relevant citation.

Earlier in this book, I provided a guide to the differences between Vancouver and Harvard referencing styles. Here I provide the output for the references that were given in that section to demonstrate the way in which these different referencing styles take up very different amounts of space. In that section, I provided some quotes from a published paper that had nine citations. Below, you can see what these nine citations look like in the two different generic outputs: Vancouver and Harvard.

## 39.1 Vancouver style

The following references follow the text written in Vancouver style in the section on citations. Note that the order of citations here corresponds to the order in which they are cited.

DOI: 10.1201/9781003212560-42

[1]Wake DB, Vredenburg VT. Are we in the midst of the sixth mass extinction? A view from the world of amphibians. *Proceedings of the National Academy of Sciences.* 2008;105(Supplement 1):11466–73.

[2]Collins JP, Crump ML, Lovejoy III TE. *Extinction in our times: global amphibian decline.* Oxford University Press; 2009.

[3]Pimm SL, Jenkins CN, Abell R, Brooks TM, Gittleman JL, Joppa LN, et al. The biodiversity of species and their rates of extinction, distribution, and protection. *Science.* 2014;344(6187):1246752.

[4]Kupferberg SJ. Bullfrog (*Rana catesbeiana*) invasion of a California river: the role of larval competition. *Ecology.* 1997;78(6):1736–51.

[5]Dufresnes C, Dubey S, Ghali K, Canestrelli D, Perrin N. Introgressive hybridization of threatened European tree frogs (*Hyla arborea*) by introduced *H. intermedia* in Western Switzerland. *Conservation Genetics.* 2015;16(6):1507–13.

[6]Berger L, Speare R, Hyatt A. *Chytrid fungi and amphibian declines: overview, implications and future directions. Declines and disappearances of Australian frogs* Environment Australia, Canberra. 1999;1999:23–33.

[7]Daszak P, Cunningham AA, Hyatt AD. Infectious disease and amphibian population declines. *Diversity and Distributions.* 2003;9(2):141–50.

[8]La Marca E, Lips KR, Lotters S, Puschendorf R, Ibanez R, Rueda-Almonacid JV, et al. Catastrophic population declines and extinctions in Neotropical harlequin frogs (Bufonidae: *Atelopus*) *Biotropica: The Journal of Biology and Conservation.* 2005;37(2):190–201.

[9]Martel A, Spitzen-van der Sluijs A, Blooi M, Bert W, Ducatelle R, Fisher MC, et al. *Batrachochytrium salamandrivorans* sp. nov. causes lethal chytridiomycosis in amphibians. *Proceedings of the National Academy of Sciences.* 2013;110(38):15325–9.

## 39.2   Harvard style

The references below are exactly the same as those above, but follow the Harvard referencing format. The order of these references is alphabetical. If a paper has the same authors the oldest paper usually comes before subsequent papers.

Berger, L., Speare, R., Hyatt, A., 1999. *Chytrid fungi and amphibian declines: overview, implications and future directions. Declines and disappearances of Australian frogs.* Environment Australia, Canberra 1999, 23–33.

Collins, J.P., Crump, M.L., Lovejoy III, T.E., 2009. *Extinction in our times: global amphibian decline.* Oxford University Press.

Daszak, P., Cunningham, A.A., Hyatt, A.D., 2003. Infectious disease and amphibian population declines. *Diversity and Distributions* 9, 141–150.

Dufresnes, C., Dubey, S., Ghali, K., Canestrelli, D., Perrin, N., 2015. Introgressive hybridization of threatened European tree frogs (*Hyla arborea*) by introduced *H. intermedia* in Western Switzerland. *Conservation Genetics* 16, 1507–1513.

Kupferberg, S.J., 1997. Bullfrog (*Rana catesbeiana*) invasion of a California river: the role of larval competition. *Ecology* 78, 1736–1751.

La Marca, E., Lips, K.R., Lotters, S., Puschendorf, R., Ibanez, R., Rueda-Almonacid, J.V., Schulte, R., Marty, C., Castro, F., Manzanilla-Puppo, J., others, 2005. Catastrophic population declines and extinctions in Neotropical harlequin frogs (Bufonidae: *Atelopus*) *Biotropica: The Journal of Biology and Conservation* 37, 190–201.

Martel, A., Spitzen-van der Sluijs, A., Blooi, M., Bert, W., Ducatelle, R., Fisher, M.C., Woeltjes, A., Bosman, W., Chiers, K., Bossuyt, F., others, 2013. *Batrachochytrium salamandrivorans* sp. nov. causes lethal chytridiomycosis in amphibians. *Proceedings of the National Academy of Sciences* 110, 15325–15329.

Pimm, S.L., Jenkins, C.N., Abell, R., Brooks, T.M., Gittleman, J.L., Joppa, L.N., Raven, P.H., Roberts, C.M., Sexton, J.O.,

2014. The biodiversity of species and their rates of extinction, distribution, and protection. *Science* 344, 1246752

Wake, D.B., Vredenburg, V.T., 2008. Are we in the midst of the sixth mass extinction? A view from the world of amphibians. *Proceedings of the National Academy of Sciences* 105, 11466–11473.

---

Be aware that these styles (Vancouver and Harvard) represent two broad approaches, every publisher has their own variation on these, and that they can vary substantially between different journals. Full details are always provided by the journal (in their **Instructions to Authors**), and you can refer to these to see how your references need to be formatted. I also provide some pointers towards using reference managers. These can save you a lot of time, if you've already invested time in setting them up. Alternatively, they can be very frustrating.

Lots of journals (or perhap's it's their editors) are very fussy about the way that citations are given in their publications. If you have used a decent reference manager then you won't care, and it'll just be a case of finding the appropriate format from their repository (or making it yourself).

There are lots of better things that you could be doing with your time instead of formatting references. And so I'm not going to spend a long time here telling you that you must do it right. I really believe that on first submission of your manuscript you should be able to submit your references in any format. it really only makes sense for you to do all the work formatting them if you aren't going to pay to publish the paper. An increasing number of journals do allow you to use any referencing style (within the generic Vancouver or Harvard format) when you first submit. You are only then required to format them if your ms is likely to be accepted.

There are lots of anal academics out there who will delight in looking for every missing full-stop, comma and capital letter missing from your references. Perhaps it's because they feel secure pointing out these errors but are incapable of knowing whether or not you've written good or bad science. Please don't join them and, if you can, look for a better world where you spend your valuable time doing more important things.

## 39.3   Digital Object Identifiers (DOIs)

A DOI[1] is an international standard (ISO[2]) unique character string to identify physical, digital or abstract objects. Their beauty is that they are persistent over time, so once issued they will always be a way to reach a particular object.

DOIs are extremely useful as you can usually click on an active link DOI and go straight to the article in question. Therefore it is well worth adding the DOI to your references if you can. Some journals will allow this (and even demand it), while others have yet to come around to how useful they are.

However, DOIs cannot replace references, otherwise, we'd need to be able to click on every link all the time, and couldn't read any paper without a connection to the internet. It's still really useful to be able to read a formatted reference at the end of a paper.

### 39.3.1   DOI tools

If you are writing your references by hand and need a DOI for every reference then there is this very useful online software that will provide the DOI if there is one for every reference you enter: doi.crossref.org[3]

There's an equally useful database that provides BibTex for DOIs that you enter: doi2bib.org[4]

---

[1] https://www.doi.org/
[2] https://www.iso.org/
[3] https://doi.crossref.org/simpleTextQuery
[4] https://doi2bib.org/

# 40

## Tables

If you think that tables are just about displaying a series of boring numbers, then it's time to free your mind. Tables are an amazing vehicle to display information that can be numeric, written or pictorial. From the simplest of two by two tables with several levels of organisation, the reason for putting your information in a table is that you can provide it in a simpler and easier way to the reader than if you were to put all the information in full into a paragraph of text. There are few rules on Tables, but you'll find some guidelines in this Table 40.1.

Most journals want tables formatted so that there are only horizontal lines. This makes them easier to print, but it can make following text across a large table difficult. Shading every other row really helps with this, and if you have the option of using this in your thesis then I'd recommend it.

As a biologist, I first started to use tables to provide large quantities of numbers. Tables of morphological measurements or meristics (counts). These were always tricky to put together in word processors until they became better integrated with spreadsheets, and now it's really easy to put these tables together and even format them in a spreadsheet and then cut and paste them into your word processor.

DOI: 10.1201/9781003212560-43

**TABLE 40.1: Some general dos and don'ts for tables.** There are exceptions for all of these, but if you can stick to them you are unlikely to go far wrong.

| Do | Don't |
|---|---|
| Use a table as a means of reducing the length of your text. | Merge cells, especially in irregular ways, as this makes formatting and copy/paste of your table prone to errors. |
| Make your headings clear and concise. | Omit units of measurement. |
| Summarise extensive data. | Try to include every data point (that's what Supp Info is for). |
| Use tables to provide repeated statistical information on tests or models. | Feel the need to include all of these in the ms (that's what Supp Info is for). |
| Provide a full table legend that explains exactly what is in it (see figures). | Make tables so large that they have to run over several pages. If this can go into Supp Info then it should. |
| Be adventurous by using tables to produces ideas, even with entire paragraphs of text or photographs. | Feel obliged to stick to using tables only for numbers. |
| Use shading for every other row. | Have rows or columns so close together that you can't tell them apart. |

# 41

## *Figures*

The way in which to visualise your data requires another book on the topic, as there are a myriad of decisions to make on all sorts of topics. What I provide below is a very brief start to thinking about these topics. If you need to explore more of the world of data visualisation, I suggest that you read books on this topic. Happily, people like Kieran Healy (2018), Jack Dougherty and Ilya Ilyankou (2021) and Claus Wilke (2019) have each written such books. Claus points out that while there are ways in which you can go wrong when plotting data (such as drawing lines to infer relationships that are not there), the skills in data visualisation are more in the aesthetic and interpretive environment of what tools to use that make the data easiest to interpret. How will the figure enhance understanding of the data and respond to your hypotheses?

Note that the principal software used for drawing data in Dougherty & Ilyankou's, Kieran's and Claus's books is R (see Part I), and the many packages that have been developed for use within the R environment (R Core Team, 2021). The distinct advantages that this produces are the almost limitless additions and details that you can make to these figures to produce exactly what you want, in a vector format. The drawback is that you might need to invest some time to learn how the graphics package works, and how to manipulate it to do what you want.

I agree with the approach of Dougherty and Ilyankou (2021) who suggest that you start any figure by mapping out what you want to convey and how. Sketch this out and discuss it with your advisor.

## 41.1 Graphs

You should be aware of all the things that make up a good graph already. Here I'm going to provide some extra pointers on what I think is important and should be considered when drawing a good graph:

- **Labels** This seems completely obvious and yet it is a common problem with

DOI: 10.1201/9781003212560-44

submitted papers. Axes should be labelled with both an explanation and the units in which they were measured.

  – **Capitals** Be consistent in capitalising labels and other words in your figure. Avoid using just capitals (e.g. ALL CAPITALS ARE DIFFICULT TO READ AND UNSIGHTLY)

- **The scale** Make sure that the numbers of the scale are large enough to read even when the figure is reduced the size that it will be printed on the page. This includes the text that crossed the axes.

  – If you have used a log scale, make sure that this is prominently labelled.

- **Data points** I am a great fan of including all the data points that go to make up any summary statistic that is displayed on the graph. Thus, if you provide a mean or something like a box plot for your x-axis category, then please add all data points as well. This can be done using the jitter function in R.

- **Don't draw relationships that aren't there.** The ease of plotting graphs makes it possible to display relationships between data points that don't exist. This is most commonly done across a set of discrete measurements when a continuous line links them without any data to suggest this relationship. Be especially careful when plotting lines that you can infer a continuous value between points plotted.

- **Compiling graphs.** If you have a series of graphs that share the same x- or y-axis then consider compiling them together. This essentially turns them into a composite figure, but it often means that readers can compare data easily from one graph to another. Remember that you will have to make sure that the scales on the shared access are all the same.

- **Use consistent symbols and colours throughout your thesis.** Your thesis will likely plot data on similar variables in many ways throughout. By using consistent symbols (e.g. circles for females and triangles for males) and colours throughout the thesis for the variables that chapters have in common, you will help your reader understand a figure much more quickly. The consistency of the graphical environment of a thesis will facilitate understanding.

  – Consistency is especially important across panels within a composite figure.

- **Colour.** The reason for all the fuss about not having colour in figures is mostly to do with the extra cost involved in printing them. Typically these days, it is possible to use colour in figures and for these to be reproduced without any extra charge online. Thus if the data that you have is best shown with colour then use colour. There are some things to be mindful of when using colour.

  – **Colour blindness:** This affects quite a large portion of the population, particularly men. To avoid using colours that can be easily confused simply look them up.

  – **Printing colour in monochrome:** some people still like to print out papers in order to read them, annotate them or share them at a journal

club. Using a monochrome printer many colours are indistinguishable. You can therefore change the hue or intensity of a colour to make sure that it can be distinguished when it is printed.
- **Aesthetic quality:** some colours don't go well together. A lot of people have already determined this and you can find a good guide here[1].

## 41.2  Maps

There are many reasons why you might need to include a map in your paper when writing for biological sciences. I find that the easiest way of creating Maps is to use GIS software like QGIS or ArcGIS. Both of these require a good degree of learning in order to master. To have a map with sampling points is relatively straightforward. Especially with QGIS you can watch 'how to' videos on the internet to learn most of what you need to know. Things to remember when you're drawing a map: If your map is a detailed view of an area, then consider a small context map in the corner showing where your area is – Include the scale – Provide a North arrow, especially if you have rotated the view – Some journals require a specific mention of the geographic projection see here[2], or to provide a citation for each layer of the map (which makes sense when you think about it). Thus it's a good policy to make sure you know the provenance of the layers that you are using. When using colour see these[3] handy sets of colours to use at Colorbrewer. Make sure that any points you have included are big enough so that you can see the point and the shape of a point when the figure is reduced – Strictly speaking maps in figures should not have embedded legends (keys). The correct way to explain symbols is to do it in the figure legend. However, sometimes if there are many different symbols it is more expedient to use an embedded legend.

### 41.2.1  Maximising the content of your figure

Sometimes you will have the option of including more information in a figure than the original plan that you started with. This is perhaps especially true when it comes to producing maps with GIS because you have the option of adding extra data layers. But remember that this information should not be gratuitous. In Figure 41.1, we can see a map that shows a number of sampling sites. The map has a scale bar and north arrow as needed.

---

[1] https://colorbrewer2.org/#type=sequential&scheme=BuGn&n=3

[2] https://www.gislounge.com/map-projection/

[3] https://colorbrewer2.org/#type=sequential&scheme=BuGn&n=3

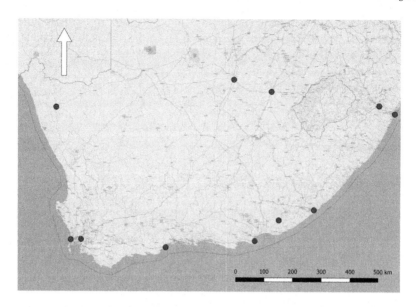

**FIGURE 41.1: A plain map can give all the information needed.**
This map adequately shows the positions of sample sites (or anything else) as
red points within the southern African region.

However, it is relatively simple to add more information to this map in the
form of layers. In Figure 41.2, a digital elevation model (DEM) allows us to see
the relative elevation at the sampling points, while transparent colours provide
information about the different biomes in which the samples were made. Lastly,
I've used a piece of sea to insert an image of the species sampled. There is
also a reference map (top left) showing the area of southern Africa where the
sampling points occur. This figure is now much more informative, and has
greater aesthetic appeal to me, but all of this added information is only of use
if relevant to the manuscript or chapter.

## 41.3  Conceptual diagram

Conceptual diagrams or figures are very useful in the introduction or methods
when you have undertaken a complex study design. There are many different
ways of doing this, but essentially the information needs to be conveyed in a
way that is more simple in the figure than writing it out. Boxes and arrows are
sufficient for most purposes. If the figure gets too complicated to understand,
it's probably of little use.

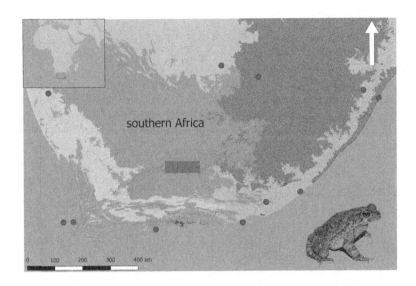

**FIGURE 41.2: Maps tend to have a lot of space, and GIS provides the opportunity to add layers to give rich information, as well as improving aesthetics.** This map shows lots more information about the biomes in which each of the sample sites (red dots) falls, the elevation and an image of the species sampled. The inset of Africa (top left) shows the area sampled with a red square. I've also managed to fit in an image of the species sampled into the sea (bottom right).

I encourage my students to draw conceptual diagrams of how the thesis chapters relate to one another, and then to include this diagram in the introduction to the thesis to help examiners and any other readers to understand how the chapters are interrelated.

I suggest that it is also very useful when you're undertaking a review to draw a conceptual diagram for yourself of what you were trying to achieve. This may or may not then be included in the review when it comes to publishing.

In order to make your conceptual diagram clear the following points should be followed:

- Keep any text in the diagram large enough so that when it is reduced to printing size it can still be read
- Try not to write full sentences but just notes
- Make use of arrows to show how ideas are related
- Look at other conceptual diagrams, especially those that resonate with you as a reader

## 41.4   The composite figure

I saw the usefulness of composite figures when I noticed other papers were including small pictures of species in phylogenies. As a reader, I really liked this way of understanding more about the animals that were being studied. The composite figure also comes about by making use of blank spaces that naturally occur within figures. Many graphs have large white spaces toward the top (left or right), and small images of study animals, cartoons of study designs or systems can be inserted to improve the immediate recognition of what the figure is portraying.

### 41.4.1   Ideas for composite figures

Include a picture of the subject species in a small corner of blank space in one of your figures. For example, maps with a large area of sea (if your subject is a terrestrial species) are especially good for this (see Figure 41.2).

## 41.5   Sending your graphics file to the publisher

One consistent problem is the lack of understanding that authors have for different file types. It can be rather confusing as these are stated in lots of different ways, and often in units that are not intuitive. Journals usually have long and detailed explanations of how to do this in their instructions to authors, yet this is one of the most common difficulties that authors have. I will provide a brief overview of the most common problems, but better overviews exist and I encourage you to read them (e.g. Wilke, 2019).

Be aware that the publisher won't be able to change anything within your figure, like the fonts. This means that you will need to use the same font in your figures as stipulated by the journal (in the instructions to authors). You also need to be critical of the size of any fonts that you use so that they can be seen even if the figure is reduced to fit the page width, or half page width. Be especially careful of numbers on scales as well as labels.

Most publishers will require you to name the files of your figures to correspond with their instructions to authors. If they don't have specific requirements, then use clear file names (e.g. Figure_1.png), and not cryptic ones. Remember that each figure needs a single file, so if you have produced a composite figure, then you will need to output this as a single graphics file. You will need to send them a generic file type (e.g. eps, jpg) and not something specific to the software that you used (e.g. ai, cdr).

### 41.5.1 Vectors vs. bitmaps

The biggest difference among image types is the difference between vectors (a line drawn between two points) and bitmaps (using pixels to create images). Vector diagrams will produce superior images on screen that will allow you to zoom in without loss of resolution (e.g. svg, eps, pdf). With a bitmap, the more you zoom, the more you will be able to see the pixels that make up the figure. If at all possible, create your figures using vector graphics (such as in R or drawing software such as Adobe Illustrator), and store them in this format. The only exceptions will be images (e.g. png, jpg, tif, raw, gif, bmp) that you add to composite figures, insert these with the highest resolution available.

If your figure is a photograph (or another tonal image type), then it will have to be a bitmap. However, if you have a line drawing, try to get it output as a vector rather than a bitmap as not only will the quality be better, but your options for editing and improving will be greater (should the editor require this).

### 41.5.2 Sizing your figure

When you export your graphic from the graphics package that you are using to draw it, specify the resolution (most often in dpi: dots per inch) given by the publisher. This only relates to bitmaps and composite figures, as vectors do not have a resolution. Minimum resolution for photos and tone images is 400 dpi, and 600 dpi for line drawings. Remember that the dpi will change if the size of your figure changes. Hence you must know the printed size of your figure (in cm or inches) on the page. This is usually either one or two columns. Rarely, a journal might allow you to use landscape instead of portrait for a figure. As a rule, you should use the page setup in your package to specify the output size of your figure (page layout function) before you start, and then set the resolution before you export the figure. It's always possible to draw at a higher resolution and convert to a lower resolution. Low-resolution images need to be redrawn to make them higher resolution.

If you are using R, you can specify your graphical output in your coding. If you are using a GUI (like R Studio)[4], you can choose to export as an image (i.e. bitmap) or pdf (i.e. vector). R Studio will allow you to set a generic or custom page size before you save.

---

[4]https://www.rstudio.com/

## 41.6   The figure legend

The most important point with the figure legend is that it should be a stand-alone explanation of what is in the figure. The reader should not have to refer to the text in order to understand what is in the figure. For example, if the figure is about a species then species name should be included in full. The explanation of any variables or categories needs to be provided. But it should not be taken to absurd levels, and is not a reason to repeat everything that's in the materials and methods. Take a look at some figure legends that are already published in the journal that you are hoping to submit to and you should have a good idea of what to write. Figure legends need to be full, grammatically correct sentences.

Remember that the figure legend is placed underneath the embedded figure in your thesis. This means that the graphic itself should not have any title. Traditional submissions to journals require figure legends to be at the end of the manuscript, before the figures themselves. Many journals now allow you to embed a lower resolution image of the figure on first submission together with the legend underneath in order to facilitate peer review. Check on the requirements for your institution or journal before finalising this.

A trend with online journals is that each figure (and table) needs a topic or title sentence, followed by a description. This is a good way forward with any figure legend. Both need to be uploaded along with the image file when you submit the manuscript.

# 42

## Who Did What?

More and more journals are asking authors to supply information about who did what. This is an attempt to increase transparency and reduce the incidence of ghost authorship.

There are a lot of politics around authorship. How do you decide whether or not someone did enough to be included on an author line. This is a very important point. All that I will say here is that this should have been decided early on in the study. Everyone should know whether or not they are going to be an author at the time that they participate. No one should ever find out but they were not an author by reading their name in the acknowledgements.

### 42.1 CRediT where it's due

After many iterations, there is now a widely recognised list of 14 different ways in which authors can contribute towards a study (Table 42.1). Even when the study is a chapter in your PhD thesis, it's unlikely that you can put your name next to every role in the CRediT taxonomy. This does give you some realistic perspective about the collaborative nature of science.

**TABLE 42.1: There is now a recognised list of different ways in which authors contribute to publications.** The CRediT taxonomy allows you to ascribe contributions from each of the authors into one or more of 14 different roles.

| Ways in Which Authors Contribute to Studies |
|---|
| Conceptualisation |
| Data curation |
| Formal Analysis |
| Funding acquisition |
| Investigation |
| Methodology |
| Project administration |

DOI: 10.1201/9781003212560-45

| Ways in which authors contribute to studies |
| --- |
| Resources |
| Software |
| Supervision |
| Validation |
| Visualisation |
| Writing – original draft |
| Writing – review and editing |

If you want to know more about each of these roles and how they are defined in the CRediT taxonomy then please visit the site. You can also 'claim' these roles under your Rescognito profile[1], linked to your ORCID account.

Even if the journal that you are submitting to does not use the credit taxonomy, I would suggest that you use this as the basis for your saying who did what in the publication. Otherwise, you could end up with what is known as hyperauthorship, a phenomenon evidenced by massive co-authorship levels (Cronin, 2001).

## 42.2   Ordering the author line up

In Biological Sciences, it is traditional for the author who did most of the work to take the first position. For you and your thesis you are most likely to be the first author, certainly for all of your thesis chapters.

The last position is normally for the head of the laboratory in which the work was undertaken. This would usually be your advisor, but could be another collaborator, if, for example, you went to do some work in their laboratory. Funding also plays a role here, and if you are unsure about who should be listed last, then ask your advisor.

In between the first and last authors is a bit of a no-man's land. Some people order it by who did most (closest to first), but then there is also a push to be closest to last place (closest to being more senior). Alphabetical also works, but that often benefits or penalises those who have names that regularly appear higher up or lower down in the alphabet. One colleague is always first to suggest alphabetical order as he regularly ends up as last author.

There is such a thing as shared first (and even last) authorship. You see these appearing with increasing regularity, marked with an asterisk or other symbol. Hence, there can be several top spots, and these should work just fine for your CV.

---

[1] https://rescognito.com/

# 43

## Supplementary Material

Twenty years ago, the idea of supplementary material for a journal article was practically unheard of. Today, it seems that few papers don't have some supplementary material associated with them. Why this big change?

## 43.1   What can go into supplementary material?

Practically anything can go into supplementary material.

- If you have lots of extra analyses that are too large to go into an article
- If your methods are extensive and there's a word limit in the journal
- If your analytical tools throw out a bunch of super interesting graphics, but you can only use one or two
- Most of all, you can put in the raw data that makes up the basis for your work, together with the analytical script that you used to analyse it

## 43.2   Towards transparency and repeatability of scientific studies

This is a really positive movement in science and it should receive all of our support as it is part of the greater scientific project (see Part II). Being proud of your work is being proud of the entire product, including the data and the analysis. If you felt they were good enough for peer review, then you should be happy to make them all freely available. Not sure if you haven't made some mistakes? We all make mistakes and you shouldn't think that you will have an entirely fault-free career. Whenever and wherever possible conduct your professional due diligence and be open and honest with your research. Science is a career when we should openly acknowledge errors and mistakes that we have made in the past. There's no problem with this. Indeed, the problems only start when you refuse to acknowledge past errors and mistakes, making you prone to motivated reasoning and confirmation bias.

There are lots of data repositories out there. Some specialise in particular kinds of data, like genetic data in GenBank[1], or sound data in The Macaulay Library[2] at the Cornell Lab of Ornithology. Others are general repositories of data and you can post anything there, including the scripts that you used to analyse the data. By doing this you ensure that any people that want to use your work in the future have full access to it. Your data deposit will be given a DOI. You can use this DOI in your publication so that readers know where to find the data.

Once I happily allowed the publishers to store my data. These days, I simply don't trust them to curate and keep my data for posterity. The way that most publishers of academic journals behave, there is no reason for them to make sure that my data is safe and secure. Therefore, I would rather go to an independent curator of data that specialises in not-for-profit data curation. One example is Zenodo[3] which is built and operated by CERN and OpenAIRE. This site has the advantage that it is integrated with GitHub (if you are already a user, but check it out if you aren't). You can choose to open your data immediately, or embargo it until your paper is published. It's a great platform, and so much more secure than using a publisher. Another example is the Open Science Framework (OSF)[4] that aims to be a one-stop shop for all of your storage needs. You will need to make sure that your data is correctly sorted with sufficient metadata for someone else to understand (see Roche et al., 2015).

Once, we did statistical analysis with a GUI and a point and click approach. Forget to click a checkbox and the results were not repeatable. As the analyses have become increasingly sophisticated with more and differing pieces of software, it is becoming increasingly important to record the code that you write that gets your results together with your data. This is the difference between making your results and, therefore, your study repeatable.

Talk to your advisor. Your lab or your institution may have a repository that they prefer.

---

[1] https://www.ncbi.nlm.nih.gov/genbank/
[2] https://www.macaulaylibrary.org/
[3] https://zenodo.org/
[4] https://osf.io/

## 43.3 Why stop with just depositing your data and analyses online?

Let's face it, you probably put a lot of effort into proposing your thesis hypotheses. Why not deposit this too? Remember that it does not have to be accessible to the world, but it would improve the integrity of your study if your original hypotheses are the same as the ones that you submit for publication. Some journals are now asking whether you have stored hypotheses ahead of conducting the experiments. They don't stop you from submitting if you haven't. But why not? It's a great step forward in terms of transparency and provides more credibility when you eventually write that manuscript with the same aims. And it shouldn't stop you from publishing another study that happened along the way, even though you hadn't planned it. I suspect that in future funders will require us to deposit our successful funding proposals in an online repository before they release funds – hopefully, they'll also be willing to support these facilities financially.

# Part IV

# After the Data Chapters

DOI: 10.1201/9781003212560-47

# 44

## Now That You Have Finished Your Data Chapters

In this last part of the book, I am going to presume that you have done all of the work for the data chapters, and written them up as distinct chapters of your thesis. When you look at PhDs that are stored in your institution's libraries, or those that you've referred to in the course of your study, you will notice that they have two additional chapters in addition to data chapters. These are often quite short (although lengths vary), and serve to introduce and conclude the PhD thesis (Figure 44.1). While these are not usually considered to be the 'meat' of the PhD itself, they are important sections to reflect on and carefully consider, as they attempt to pull together data chapters that are written as stand-alone documents into a cohesive thesis. To this end, they require the same careful planning that any other section in your thesis deserves, with one important condition: that they should not impede the timely submission of your thesis.

**FIGURE 44.1: A typical construction of the data chapters in your thesis together with an introduction and conclusion.** The exact construction of your thesis may vary, especially with the number of data chapters that you have. The relationship between thesis chapters may also not be linear, as presented here.

At this point, we have to acknowledge that different institutions require different deadlines of their students in order to finish their PhD studies and submit their theses for examination. In most institutions (to my knowledge) the time to submission is the most important, and can vary between three and five years (most institutions stipulate a minimum period of two years and some have no maximum duration). This period is critical with respect to funding. If funding ceases and your institution allows you to continue your studies, you may have to raise the funding to pay fees as well as to maintain your living

DOI: 10.1201/9781003212560-48

costs. After submission, there will be a period of time while your thesis is in examination, and another period before the final graduation. Each of these milestones comes with its own set of deadlines and conditions. It is well worth being aware of all these issues well in advance of any submission deadline (ask your advisor). Important aspects for you to note are whether or not you need to remain registered at your institution after you have submitted your thesis, and whether or not your institution will require payment for your registration.

Whatever your deadlines are will dictate how much time (and energy) you have to devote to the opening and closing chapters of your thesis. They can become publishable in their own right, or at the bare minimum, they can serve to allow you to submit a completed thesis. Here I will focus on the latter, minimum requirement (in my eyes), with the understanding that you should dedicate as much time as you have available to produce a quality product that you can feel proud of.

The next chapters deal with writing the introductory chapter, the concluding chapter, and then finally formatting your thesis for submission. These final hurdles require some time and thought. Given that not all of you will have all of the time that you need, this is often the time when the demand on your writing skills reaches a peak. If you are in this position, I urge you to use your time wisely and to plan carefully what you will write.

# 45

## How to Introduce Your PhD Chapters

For each of your data chapters, you will have already written an introduction section, and the same principles that you used to construct this will apply here. Instead of aiming to introduce any one of the hypotheses in the chapters, however, your introduction will aim to introduce the ideas and concepts explored inside the data chapters of the thesis.

My suggestion would be for you to start with the conceptual diagram of the thesis that you produced in your proposal. If you didn't construct one then, you may be reluctant to do it now, but having a good conceptual idea of how the components of your thesis fit together will really aid you in putting together both this introduction and the final conclusion. For your proposal, you may also have written a general introduction, and it would be worth revisiting this now, even if you don't actually use any of it.

Once you have a conceptual diagram of your thesis, my suggestion is that you formalise this into a diagram that you can present in the introductory chapter. Even if it's a very simple concept (like Figure 44.1), it will help your examiners (and any other readers) understand what to expect in your chapters and how they are interrelated.

Next, produce a hierarchy of hypotheses for all of the chapters and concepts in your thesis (see Figure 8.2). This will help you to make sure that you cover all of the big ideas and concepts in the outline. Your outline should follow the standard funnel formula (see Figure 34.1), this time with the highest level ideas in your hierarchy of hypotheses given full prominence in the first paragraph.

Although it's tempting, I'd urge you not to copy any text from any of the introductions in any of your chapters. By all means, draw upon their contents and references therein to give you inspiration, but this introductory chapter can't be a cut and paste job (Figure 22.1). The reason being that your examiners will spot this straight away, and it won't put your work in the best light. As before, if you can produce an outline of what you need to say in your introductory chapter, you should be well practised enough (by now) to write original text that will do the job of introducing all of the chapters.

DOI: 10.1201/9781003212560-49

## 45.1   Citing your chapters

Your aim in this introduction is to introduce all of the chapters in your thesis, thus you will to cite them. My suggestion would be that you leave these citations to the final paragraphs (or subsection) of the introduction, and not have a structure that centres around introducing each chapter as a paragraph. Remember that you may well have this (in the form of abstracts) throughout the thesis, so it would be redundant to do that here. Instead, try to introduce chapters by concentrating on their themes and/or contributions to the bigger ideas that you address in your work.

The way in which to cite your chapters is most straightforward here by using their chapter number (i.e. **Chapter n**). This will make it obvious to the examiner (or any other reader) that it is within the content of the thesis. If chapters are published, you could add a citation to the paper in addition but not instead of.

## 45.2   Formulate the objectives of your study?

At this point in your thesis writing, it is worth looking back to the criteria used to judge a PhD at your institution. Yours may differ from those stated in this book, but it is clear that you should be able to formulate the objectives of your study, and that this includes not only the objectives of the individual data chapters, but of the study overall (here in the introductory chapter).

By the end of the introductory chapter, anyone reading should know what the thesis is about, and what it is that you aim to find out. This doesn't mean that they know what you found in each of the data chapters, but that they know the topics explored within the thesis and what was done to address them.

## 45.3   Review consistent thesis themes

If your thesis has a study organism, experimental approach, or environmental system, then it would be appropriate to provide a reasonably comprehensive review of this in the introduction. The aim of this would be to explain why this theme was chosen. The extent of the review would depend entirely on how much time and energy you have to devote to this.

Ideally, you would aim for your first chapter to be a review of the literature in your study area, with the last section of the review rationalising the need for the particular approach that you have taken to address unexplored areas. However, as stated above, it may well be that you don't have the time to write a comprehensive review for your introductory chapter. If you do, then the most important aspect will be what to cover in the review (exactly what portion of the literature), and whether and how you want to publish it. Publications of reviews broadly fall into traditional reviews and systematic reviews (including meta-analyses). The latter require considerable amount of time and effort, and systematic reviews (i.e. those in which you assess an unbiased slice of literature on a topic) might be the only way to publish a review in some journals or subdisciplines.

Although the review of consistent themes could be anything from two pages to an entire manuscript, you should make it clear to the examiner (or any other reader) why, out of all potential options, you have taken the particular approach you have.

## 45.4   Removing the contents of chapters to place in the introduction

You may decide that some of the contents of (particularly data chapter 1) might be more appropriate if it was moved to the introductory chapter. Before you do this I'd suggest that you discuss it with your advisor. Each of the data chapters should have already been signed off (by your advisor) and so it would be a mistake to start chopping them now without consultation. It is also perfectly legitimate to refer in the introduction to a section of the thesis where a particular topic is dealt with in detail. If you are unsure, you'd best be advised by your advisor.

## 45.5   How long should the introductory chapter be?

This is probably the most common qestion I get asked, most often by students that have only days left in which to write it. Obviously, there is no simple answer (just as there is no particular length required for a PhD). Typically, I would suggest that the introductory chapter be longer than any of the intro-ductions to any of the data chapters within your thesis, but not longer than the data chapters themselves - unless you've managed to pull off a stand-alone review.

# 46

## How to Conclude Your PhD

By the time you come to write your concluding chapter of your PhD, you should have written all of the data chapters and the introductory chapter. Firstly, I should congratulate you for getting this far. You've almost finished, and there's just one final push required to complete the document. You are now the best placed person to tell yourself what should be in the concluding chapter.

## 46.1    Using the funnel structure

In principle, you can use the same funnel structure that we described for the discussion as a way of structuring your concluding chapter (see Figure 37.1). This would include starting with a synopsis of what your thesis is trying to achieve. Clearly, this starting section may be more than a paragraph (as it was in the data chapters), and the precise structure will be dictated by the complexity of your thesis structure (Figure 44.1).

## 46.2    What to include in the concluding chapter?

I suggest that you take your conceptual outline (from the introductory chapter), and use this to thinking about how best to construct the outline of your concluding chapter. If your data chapters are used to feed into a final output, then consider using the approach of a logical argument as the outline.

The main body should review the concluding findings of each of the data chapters, and especially the way in which they interact – as these are the aspects that are least likely to have been touched on already in the individual data chapters. It is hoped that the thesis represents a body of work, and not a series of unconnected works. Thus your task in this final chapter is to present to the examiner (and any other reader) exactly how the chapters fit together into a body of work.

Your concluding chapter should provide all of the highlights and insights that have been gained from your PhD thesis. It is unlikely that you need to dwell on individuals caveats from the chapters unless there are general caveats from your approach that span chapters – in which case, these probably deserve attention.

Your concluding chapter will cite the data chapters of your thesis quite heavily, as it essentially pulls them all together, along with the objectives that they aimed to achieve, in the overarching concepts of your study area.

You also have an opportunity to point to where, in your opionion, the direction of your particular field should move next. **Do provide your opinion**. Remember that you are now one of the most qualified people in this particular area to be pronouncing on what should be done. Your opinion is therefore valid, and your concluding chapter is a good place to provide a reasoned argument about your own ideas.

## 46.3   Concluding section

Your PhD may not have one overriding simple conclusion, indeed it will likely not have, so don't try to force one if there isn't one. Instead, attempt to sum up your findings in a way that explains how they have added to and changed the existing body of knowledge in your subject area.

# 47

## Formatting Your Thesis

Let's face it, after spending so much time writing your thesis, having it formatted so that it looks very smart for submission is gratifying.

If you chose Bookdown[1] or thesisdown[2] to write your thesis in, then you'll simply be able to download the correct style file (cls) and bibliography formatting file (bib) and press the build book button in RStudio[3] and you'll have your formatted thesis in the time that it takes your computer to render the pdf. You'll still need to look through it and make sure that the output is what you want.

For other, mere mortals, using word processing packages, you will likely need to invest a lot more time trying to pull together the chapters of your thesis into the correct format. (I'd suggest that right now is not the time to learn Bookdown - unless you are way ahead of schedule!) Try to make it a fun event by listening to your favourite music (very loudly) or having some of your banned tipple (do both if you dare!). It doesn't require the same level of creative concentration required when writing your thesis, but it does need great attention to detail.

Be patient and don't try to build mega-big files in word processing documents, as they tend to crash. Instead, copy the template file and compile one chapter at a time, placing them together only at the end for a final build.

Your university will have its own stipulation for exactly how they want the thesis formatted. Here, I provide a check list for the last things to do before you submit. The chances are that your university will have a bunch of helpful links. Make sure that you follow the guidelines of your own institution well in advance of submitting your thesis.

---

[1] bookdown.org

[2] thesisdown.org

[3] www.rstudio.com

DOI: 10.1201/9781003212560-51

## 47.1   Must-do checklist before submitting your thesis

The following points are based on what may irritate or annoy your examiner
(and you really don't want to do that!):

- Spell check – yes, it sounds obvious but doing a final spell check is a good
  idea. Not only this but take the time to have your word processor ignore
  or add all of the special words (e.g. species or site names) that it doesn't
  otherwise recognise. This will ensure their consistency throughout (within
  and between chapters).
- Make sure that your language settings are set to 'UK English' (or the English
  setting for where you are based).
- Look out especially for words that have different accepted spellings like those
  ending in -ise or -ize. Decide which you want and be consistent. Consistency
  is king!
- Capitalisation of common names, place names and not adjectives. For example, 'South Africa' has two capitals, but 'southern Africa' only has one.
- Grammar check - always good to take a final look, especially for chapters
  that you wrote some time ago.
- Use the word processor automated options to help you.
- Have your computer or another device read the text so that you can hear
  anything obviously wrong.
- Pay attention if you have used 'we' or 'I' and make sure they are consistent
  in your thesis. As a rule I encourage 'I' in a thesis unless the chapter is also
  a manuscript, in which case 'we' is correct if there are multiple authors.
- Page layout. Really important to get this right in your template. Make sure
  that your template has:
  - Correct paper size (A4 and not US letter - or visa versa!)
  - Margins
  - Line spacing
  - Page numbers
  - Line numbers (really helps your examiners)
- Headers and Footers. If you can manage a chapter-specific header, it's useful
  to show your name and a short chapter title.
- Sections and subheadings. I've encouraged you to use subheadings through-
  out your thesis. Here you have a chance to number them sequentially. This is
  very useful for your examiners and may be a requirement for the university.
  Using the word processor's built-in functions will make this task consistent
  and easy.
- I dislike writing within a formatted document (as word processors can start
  getting weird), so my preference is to cut and paste written text into a
  template at the end.

- Remember to give them a check through before handing it in. If you've done the sections correctly, then the contents page will come out correctly.
- Title page - prescribed very strictly by the university. The librarians place a watermark after final submission to the library. It may be tempting to change your title now that you know what's in the thesis, but many universities have strict policies about this. If you want to change the title, make sure that you are able to do so.
- Content page – word processors can do this automatically if your thesis is formatted correctly throughout (see sections and subheadings above).
  - You can check this to make sure that you've done all of your sections and subheadings correctly.
- Acknowledgements – this is your time to say thank you to all the special people that have helped during your study. There are probably more than you realise, but in addition to your friends and family (who most people don't forget), think about the people who administered the work, lab mates (past and present) who were always there to help, and people who gave permission at study sites.
- References – probably one of the most dreaded sections of any thesis preparation, but they do have to be done. If you're one of these people that has everything in a database, then you'll be laughing or cursing your database throughout. While it might be tempting to only look through the data within the database, spend some time to see how it's displayed in the thesis. A mistake in the reference database will be multiplied many times in the thesis. Remember that examiners love to take a random look through the reference section to make sure that it's all good. After years of painfully entering references themselves, they know just what to look for.

## 47.2   Mistakes people make

Other than the obvious things, all mentioned above, here are some of the mistakes I've seen.

- Submitting the wrong version (yes, this does happen!). Probably worse if having a mixture of right and wrong versions for different chapters (worse because it takes longer to sort out)
- Last-minute additions to text with incorrect spelling and or grammar
- Two correctly spelled synonyms sitting next to each other when only one is desired (probably came about when editing)
- Forgetting to check for plagiarism
- Comments and or edited text (especially when it's marked as being by someone other than the student).

- Page numbers that start again and again at different sections
- Lots of blank pages or spaces (avoid blank pages if you can, and try to limit the amount of blank space (never >half a page).
- Leaving important people out of the acknowledgements (e.g. advisor, administrators, funders, etc.)

# The Obligation to Publish Your Work

Now that you've submitted your PhD, you still need need to do something with it. For most of you this will be publishing it in a journal of your choice. Publication is important. Here is a list of reasons why you have some obligation to publish the work in your thesis:

- If your work is unpublished, then it will not be used. Without use, all the work that you put into it falls away.
- Your published work will become the foundation for future students and academics who are active in your field. This might well be in another country or on another continent.
- Your work was most likely paid for by taxpayers in the country where you studied. Publishing this work is a way of handing back the value of what you found. If it is published Open Access, the public can also read it for themselves.
- Other funders of your thesis work may have made publication a prerequisite of the funding criteria.
- It is hoped that by publishing your work it will become more accessible to the scientific community.
- By publishing your work you will find that both you, and your work, become known by an international community that may well invite you to participate in the academic process further (Marks et al., 2013).
- Both inside and outside of science as a career, peer-reviewed published papers are seen as an important accomplishment in addition to the production of your thesis.
- The work that you have already put into your thesis (especially if you have followed the advice in this book!) will mean that you are close to having chapters that can be submitted as publications.
- Your advisor and lab colleagues who helped with your thesis work may depend on your publications.
- There is a lot of satisfaction to be had by seeing your work published. It has been argued that this depends on who you are and where you come from (Husemann et al., 2017).
- The satisfaction of having someone say how enjoyable or inspiring your work is to read will certainly not be as frequent as if you managed to publish a

novel or newspaper column. But when it does happen, it will bring a smile to your face.

This is by no means an exhaustive list, but I hope it will give you some insight into the importance of publishing your results. To counter this assertion, please be aware that there is also research that suggests that publication driven science is not healthy, and that we need a new paradigm to motivate science (e.g. Stergiou and Lessenich, 2014). Indeed, it has been suggested that the 'publish or perish' mentality of academia has resulted in the retention (and even deterioration) of poor experimental design, and hence higher false positives in the behavioural sciences (Smaldino and McElreath, 2016). However, before you can join the debate about the future, I'd argue that you need to provide your credentials for the present.

Taking your work to the next level is not a trivial step, even though you already have your chapters written in the style of manuscripts. The challenges of publishing work, like other challenges in academia, are much easier if you possess inside knowledge of the system. The demystification of the next steps along the road in academia is the subject of another book.

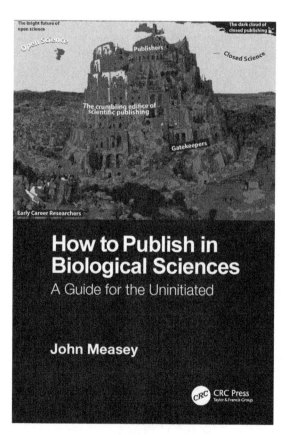

# Last Note

I really hope that this book has been helpful and that it has achieved what I set out to do: provide you with the guide on how to write a PhD in biological sciences. If you feel that this book has important items missing, is out-of-date or simply wrong, then please help. Any good guide relies upon the people that use it to keep it viable.

If you have found this book useful, then please remember that you can contribute to the project at the Github pages for this book, using bookdown. If your contributions, no matter how small, improve the book, then you will be improving the experience of doing a PhD in Biological Sciences for future PhD students, and that has to be worth something.

Remember that there is another book that starts off where this book ends: from thesis chapters to published papers. It also tackles some of the broader issues within academia, which are likely to be relevant if you are planning your career in this area.

**If I can emphasise one really important aspect of all of your PhD studies (including the writing), that is to enjoy them!**

# Bibliography

Abbott, A. (2004). *Junior Biologists Score Partial Victory over Lab Conditions*. Nature Publishing Group.

Adams, J. (2012). The rise of research networks. *Nature*, 490(7420):335–336.

Adams, D. (2017). *The Ultimate Hitchhiker's Guide to the Galaxy: The Complete Trilogy in Five Parts*, volume 6. Pan Macmillan.

Afonso, A. (2014). How academia resembles a drug gang. https://blogs.lse.ac.uk/impactofsocialsciences/2013/12/11/how-academia-resembles-a-drug-gang/.

Allen, G. E. and Baker, J. J. (2017). *Scientific Process and Social Issues in Biology Education*. Springer Texts in Education. Springer International Publishing, Cham.

Angler, M. W. (2020). *Telling Science Stories: Reporting, Crafting and Editing for Journalists and Scientists*. Routledge.

Bacon, F. (1620). *Novum Organum*. Ellis and J. Spedding (transl.).

Baker, M. (2016). 1,500 scientists lift the lid on reproducibility: Nature News & Comment. *Nature*, 533:452–454.

Becker, F. S., Tolley, K. A., Measey, G. J., and Altwegg, R. (2018). Extreme climate-induced life-history plasticity in an amphibian. *The American Naturalist*, 191(2):250–258.

Berger, L., Speare, R., and Hyatt, A. (1999). Chytrid fungi and amphibian declines: overview, implications and future directions. *Declines and disappearances of Australian frogs. Environment Australia, Canberra*, 1999:23–33.

Bonney, R., Shirk, J. L., Phillips, T. B., Wiggins, A., Ballard, H. L., Miller-Rushing, A. J., and Parrish, J. K. (2014). Next steps for citizen science. *Science*, 343(6178):1436–1437.

Bouton, M. E. (1993). Context, time, and memory retrieval in the interference paradigms of Pavlovian learning. *Psychological Bulletin*, 114(1):80.

Britton, J., Buscha, F., Dickson, M., Erve, L. V. d., Vignoles, A., Walker, I., Waltmann, B., and Zhu, Y. (2020). The earnings returns to postgraduate degrees in the UK. Technical report.

Bump, J. K., Tischler, K. B., Schrank, A. J., Peterson, R. O., and Vucetich, J. A. (2009). Large herbivores and aquatic–terrestrial links in southern boreal forests. *Journal of Animal Ecology*, 78(2):338–345.

Catford, J. A., Jansson, R., and Nilsson, C. (2009). Reducing redundancy in invasion ecology by integrating hypotheses into a single theoretical framework. *Diversity and Distributions*, 15(1):22–40.

Clance, P. R. and Imes, S. A. (1978). The impostor phenomenon in high achieving women: Dynamics and therapeutic intervention. *Psychotherapy: Theory, Research & Practice*, 15(3):241–247.

Collins, J. P., Crump, M. L., and Lovejoy III, T. E. (2009). *Extinction in Our Times: Global Amphibian Decline*. Oxford University Press.

Cope, B., Kalantzis, M., Abd-El-Khalick, F., and Bagley, E. (2013). Science in writing: Learning scientific argument in principle and practice. *E-learning and Digital Media*, 10(4):420–441.

Cronin, B. (2001). Hyperauthorship: A postmodern perversion or evidence of a structural shift in scholarly communication practices? *Journal of the American Society for Information Science and Technology*, 52(7):558–569.

Culumber, Z. W., Anaya-Rojas, J. M., Booker, W. W., Hooks, A. P., Lange, E. C., Pluer, B., Ramirez-Bullon, N., and Travis, J. (2019). Widespread biases in ecological and evolutionary studies. *BioScience*, 69(8):631–640.

Daszak, P., Cunningham, A. A., and Hyatt, A. D. (2003). Infectious disease and amphibian population declines. *Diversity and Distributions*, 9(2):141–150.

Daudin, F.-M. (1802). *Histoire naturelle des rainettes, des grenouilles et des crapauds. Avec planches*. Levrault.

Dawkins, R. (2004). *A Devil's Chaplain: Reflections on Hope, Lies, Science, and Love*. Mariner Books.

Delgado Lopez-Cozar, E., Robinson-Garcia, N., and Torres-Salinas, D. (2014). The Google Scholar experiment: How to index false papers and manipulate bibliometric indicators. *Journal of the Association for Information Science and Technology*, 65(3):446–454.

Dhar, P. K. and Giuliani, A. (2010). Laws of biology: Why so few? *Systems and Synthetic Biology*, 4(1):7–13.

Dickerson, D. (2019). How I overcame impostor syndrome after leaving academia. *Nature*, 574(7779):588–588. tex.copyright: 2021 Nature.

Dougherty, J. and Ilyankou, I. (2021). *Hands-On Data Visualization*. O'Reilly Media, Inc.

Dufresnes, C., Dubey, S., Ghali, K., Canestrelli, D., and Perrin, N. (2015). Introgressive hybridization of threatened European tree frogs (Hyla arborea) by introduced H. intermedia in Western Switzerland. *Conservation genetics*, 16(6):1507–1513.

Dumeril, C. and Bibron, G. (1841). *Erpetologie generale ou histoire naturelle complete des reptiles: comprenant l'histoire generale des batraciens, et la description des cinquante-deux genres et des cent soixante-trois especes des deux premiers sous-ordres: les peromeles ...* Imp. de Fain et Thunot.

Dunleavy, P. (2003) *Authoring a PhD*. 1st ed. Palgrave Study Skills. London: Palgrave https://doi.org/10.1007/978-0-230-80208-7.

Dunleavy, P. (2017). Citations are more than merely assigning credit – their inclusion (or not) conditions how colleagues regard and evaluate your work.

Enfield, N. (2018). Our job as scientists is to find the truth. But we must also be storytellers. *The Guardian*. https://www.theguardian.com/commentisfree/2018/jul/20/our-job-as-scientists-is-to-find-the-truth-but-we-must-also-be-storytellers.

Fanelli, D. (2010). 'Positive' results increase down the hierarchy of the sciences. *PloS one*, 5(4):e10068.

Federative Committee on Anatomical Terminology. 1998. *Terminologia Anatomica: International Anatomical Terminology*. Georg Thieme Verlag, Stuttgart.

Federative International Committee on Anatomical Terminology. 2008. *Terminologia Histologica: International Terms for Human Cytology and Histology*. Wolters Kluwer/Lippincott Williams & Wilkins, Philadelphia, PA.

Federative International Programme for Anatomical Terminology. 2013. *Terminologia Embryologica: International Embryological Terminology*. Georg Thieme Verlag, Stuttgart, 2nd edition.

Federative International Programme for Anatomical Terminology. 2017. *Terminologia Neuroanatomica: International Neuroanatomical Terminology*. Thieme, Stuttgart. https://fipat.library.dal.ca/tna/.

Firth, K. (2017). Writing a PhD in your second language: seven reasons you're doing great and five ways to do even better.

Forstmeier, W., Wagenmakers, E.-J., and Parker, T. H. (2017). Detecting and avoiding likely false-positive findings – a practical guide. *Biological Reviews*, 92(4):1941–1968.

Freeman, P. and Robbins, A. (2006). The Publishing Gap Between Rich and Poor: the Focus of AuthorAID. *Journal of Public Health Policy*, 27(2):196–203.

Frued, S. (1930). *Das Unbehagen in der Kultur*. Viena: Internationaler Psychoanalytischer Verlag.

Frost, D. R. (2020). Amphibian species of the world: An online reference, version 5.4. Type: Database.

Frost, D. R., Grant, T., Faivovich, J., Bain, R. H., Haas, A., Haddad, C. F., De Sa, R. O., Channing, A., Wilkinson, M., Donnellan, S. C., Raxworthy, C. J., Campbell, J. A., Blotto, B. L., Moler, P., Drewes, R. C., Nussbaum, R. A., Lynch, J. D., Green, D. M., and Wheeler, W. C. (2006). The amphibian tree of life. *Bulletin of the American Museum of Natural History*, 2006(297):1–291.

Galef, J. (2021). *The Scout Mindset: Why Some People See Things Clearly and Others Don't*. Portfolio.

Geraldi, J. (2021). Self-Plagiarism in Project Studies: A Call for Action and Reflection. *Project Management Journal*, 52(2):119–126.

Gerber, M., Brand, S., Herrmann, C., Colledge, F., Holsboer-Trachsler, E., and Pühse, U. (2014). Increased objectively assessed vigorous-intensity exercise is associated with reduced stress, increased mental health and good objective and subjective sleep in young adults. *Physiology & Behavior*, 135:17–24.

Ghrouz, A. K., Noohu, M. M., Dilshad Manzar, M., Warren Spence, D., Ba-Hammam, A. S., and Pandi-Perumal, S. R. (2019). Physical activity and sleep quality in relation to mental health among college students. *Sleep and Breathing*, 23(2):627–634.

Grasdalsmoen, M., Eriksen, H. R., Lonning, K. J., and Sivertsen, B. (2020). Physical exercise, mental health problems, and suicide attempts in university students. *BMC Psychiatry*, 20(1):1–11. tex.copyright: 2020 The Author(s).

Hall, S, Moskovitz, C., and Pemberton, M. (2021) Understanding text recycling. A guide for researchers. *Text Recycling Research Project*. https://textrecycling.org/resources/.

Harwood, N. (2009). An interview-based study of the functions of citations in academic writing across two disciplines. *Journal of Pragmatics*, 41(3):497–518.

Harzing, A. W. (2007). Publish or Perish. https://harzing.com/resources/publish-or-perish.

Healy, K. (2018). *Data Visualization: A Practical Introduction*. Princeton University Press.

Heger, T. and Jeschke, J. M. (2018). The hierarchy-of-hypotheses approach updated: A toolbox for structuring and analysing theory, research, and evidence. In Jeschke, J. M. and Heger, T., editors, *Invasion Biology:* Frued, S. (1930). *Das Unbehagen in der Kultur.* Viena: Internationaler Psychoanalytischer Verlag. *Hypotheses and evidence,* number 9 in CABI INVASIVES SERIES, pages 38–46. CAB International, Wallingford, UK.

Hess, D. R. (2004). How to write an effective discussion. *Respiratory Care,* 49(10):1238–1241.

Hirsch, J. E. (2005). An index to quantify an individual's scientific research output. *Proceedings of the National Academy of Sciences,* 102(46):16569–16572.

Hotaling, S. (2018). Publishing papers while keeping everything in balance: Practical advice for a productive graduate school experience. *Ideas in Ecology and Evolution,* 11:35–46. tex.copyright: Copyright (c) 2018 Scott Hotaling.

Hotaling, S. (2020). Simple rules for concise scientific writing. *Limnology & Oceanography Letters,* 5(6):379–383.

Husemann, M., Rogers, R., Meyer, S., and Habel, J. C. (2017). "Publicationism" and scientists' satisfaction depend on gender, career stage and the wider academic system. *Palgrave Communications,* 3(1):1–10.

Hutchins, H. M. and Rainbolt, H. (2017). What triggers impostor phenomenon among academic faculty? A critical incident study exploring antecedents, coping, and development opportunities. *Human Resource Development International,* 20(3):194–214.

Ioannidis, J. P. (2005). Why most published research findings are false. *PLoS medicine,* 2(8):e124.

Jenicek, M. (2006). How to read, understand, and write 'Discussion' sections in medical articles. An exercise in critical thinking. *Medical Science Monitor,* 12(6):28–36.

Katz, Y. (2013). Against storytelling of scientific results. *Nature Methods,* 10(11):1045–1045.

Kerr, N. L. (1998). HARKing: Hypothesizing after the results are known. *Personality and Social Psychology review,* 2(3):196–217.

Kupferberg, S. J. (1997). Bullfrog (Rana catesbeiana) invasion of a California river: The role of larval competition. *Ecology,* 78(6):1736–1751.

La Marca, E., Lips, K. R., Lotters, S., Puschendorf, R., Ibanez, R., Rueda-Almonacid, J. V., Schulte, R., Marty, C., Castro, F., Manzanilla-Puppo, J., and others (2005). Catastrophic population declines and extinctions in Neotropical harlequin frogs (Bufonidae: Atelopus) 1. *Biotropica: The Journal of Biology and Conservation*, 37(2):190–201.

Lawton, J. H. (1999). Are There General Laws in Ecology? *Oikos*, 84(2):177–192.

Levecque, K., Anseel, F., De Beuckelaer, A., Van der Heyden, J., and Gisle, L. (2017). Work organization and mental health problems in PhD students. *Research Policy*, 46(4):868–879.

Linnaeus, C. (1753). *Species plantarum*. Laurentius Salvius.

Linnaeus, C. (1758). *Systema naturae*, volume 1. Stockholm Laurentii Salvii.

Louw, H. (2017). Defining plagiarism: Student and staff perceptions of a grey concept. *South African Journal of Higher Education*, 31(5):116–135. tex.copyright: Copyright (c) 2017 Henk Louw.

Malaga-Trillo, E. and Gerlach, G. (2004). Meyer case poses a challenge to the system. *Nature*, 431(7008):505–506.

Marks, M. S., Marsh, M., Schroer, T. A., and Stevens, T. H. (2013). Misuse of journal impact factors in scientific assessment. *Traffic*, 14:611–612.

Martel, A., Spitzen-van der Sluijs, A., Blooi, M., Bert, W., Ducatelle, R., Fisher, M. C., Woeltjes, A., Bosman, W., Chiers, K., Bossuyt, F., and others (2013). Batrachochytrium salamandrivorans sp. nov. causes lethal chytrid-iomycosis in amphibians. *Proceedings of the National Academy of Sciences*, 110(38):15325–15329.

Martin-Martin, A., Orduna-Malea, E., Thelwall, M., and Lopez-Cozar, E. D. (2018). Google Scholar, Web of Science, and Scopus: A systematic comparison of citations in 252 subject categories. *Journal of Informetrics*, 12(4):1160–1177.

McGill, B. (2016). The 5 pivotal paragraphs in a paper. https:// dynamicecology.wordpress.com/2016/02/24/the-5-pivotal-paragraphs-in-a-paper/.

Measey, G., Vimercati, G., De Villiers, F., Mokhatla, M., Davies, S., Thorp, C., Rebelo, A., and Kumschick, S. (2016). A global assessment of alien amphibian impacts in a formal framework. *Diversity and Distributions*, 22(9):970–981.

Measey, J., Wagener, C., Mohanty, N. P., Baxter-Gilbert, J., and Pienaar, E. F. (2020). The cost and complexity of assessing impact. *NeoBiota*, 62:279.

Mueller, P. A. and Oppenheimer, D. M. (2014). The pen is mightier than the keyboard: Advantages of longhand over laptop note taking. *Psychological Science*, 25(6):1159–1168.

Murray, R. (2011). Skillful writing of an awful research paper. *Analytical Chemistry*, 83(3):633–633.

Nosek, B. A., Ebersole, C. R., DeHaven, A. C., and Mellor, D. T. (2018). The preregistration revolution. *Proceedings of the National Academy of Sciences*, 115(11):2600–2606.

Nuñez, M. A. and Amano, T. (2021). Monolingual searches can limit and bias results in global literature reviews. *Nature Ecology & Evolution*, 5(3):264–264. tex.copyright: 2021 The Author(s), under exclusive licence to Springer Nature Limited.

Ohler, A. and Dubois, A. (2016). The identity of the South African toad Sclerophrys capensis Tschudi, 1838 (Amphibia, Anura). *PeerJ*, 4:e1553.

Pandey, S., Pandey, S., Dwivedi, S., Pandey, D., Mishra, H., and Mahapatra, S. (2020). Methods of various citing and referencing style: Fundamentals for early career researchers. *Publishing Research Quarterly*, 36(2):243–253.

Perneger, T. V. (2010). Citation analysis of identical consensus statements revealed journal-related bias. *Journal of Clinical Epidemiology*, 63(6):660–664.

Pimm, S. L., Jenkins, C. N., Abell, R., Brooks, T. M., Gittleman, J. L., Joppa, L. N., Raven, P. H., Roberts, C. M., and Sexton, J. O. (2014). The biodiversity of species and their rates of extinction, distribution, and protection. *Science*, 344(6187):1246752.

Pinker, S. (2014). Why Academics' Writing Stinks. https://www.chronicle.com/article/why-academics-stink-at-writing/.

Plaxco, K. W. (2010). The art of writing science. *Protein Science*, 19(12):2261.

Popper, K. (2005). *The Logic of Scientific Discovery*. Routledge.

R Core Team (2021). R: A language and environment for statistical computing. tex.address: Vienna, Austria.

Roche, D. G., Kruuk, L. E., Lanfear, R., and Binning, S. A. (2015). Public data archiving in ecology and evolution: How well are we doing? *PLoS biology*, 13(11):e1002295.

Rubin, M. (2020). Does preregistration improve the credibility of research findings? *arXiv preprint arXiv:2010.10513*.

Sagan, C. (2011). *The Demon-Haunted World: Science as a Candle in the dark*. Ballantine Books.

Şanlı, Ö., Erdem, S., and Tefik, T. (2013). How to write a discussion section? *Turkish Journal of Urology*, 39(Suppl 1):20.

Scheel, A. M. (2020). Registered Reports: A process to safeguard high-quality evidence. *Quality of Life Research*, 29(12):3181–3182.

Schloss, P. D. (2018). The Riffomonas Reproducible Research Tutorial Series. *Journal of Open Source Education*, 1(3):13.

Shashok, K. (2008). Content and communication: How can peer review provide helpful feedback about the writing? *BMC Medical Research Methodology*, 8(1):1–9.

Silvertown, J. (2009). A new dawn for citizen science. *Trends in Ecology & Evolution*, 24(9):467–471.

Simmons, J. P., Nelson, L. D., and Simonsohn, U. (2011). False-positive psychology: Undisclosed flexibility in data collection and analysis allows presenting anything as significant. *Psychological Science*, 22(11):1359–1366.

Smaldino, P. E. and McElreath, R. (2016). The natural selection of bad science. *Royal Society Open Science*, 3(9):160384.

Statzner, B. and Resh, V. H. (2010). Negative changes in the scientific publication process in ecology: Potential causes and consequences. *Freshwater Biology*, 55(12):2639–2653.

Stergiou, K. I. and Lessenich, S. (2014). On impact factors and university rankings: From birth to boycott. *Ethics in Science and Environmental Politics*, 13(2):101–111.

Stubb, J., Pyhältö, K., and Lonka, K. (2011). Balancing between inspiration and exhaustion: PhD students' experienced socio-psychological well-being. *Studies in Continuing Education*, 33(1):33–50.

Sullivan, I., DeHaven, A., and Mellor, D. (2019). Open and reproducible research on open science framework. *Current Protocols Essential Laboratory Techniques*, 18(1):e32.

Taşkin, Z., Doğan, G., Kulczycki, E., and Zuccala, A. A. (2020). Long read | Science needs to inform the public. That can't be done solely in English. https://blogs.lse.ac.uk/covid19/2020/06/18/long-read-science-needs-to-inform-the-public-that-cant-be-done-solely-in-english/.

Teixeira da Silva, J., Al-Khatib, A., and Dobranszki, J. (2017). Fortifying the corrective nature of post-publication peer review: identifying weaknesses, use of journal clubs, and rewarding conscientious behavior. *Science & Engineering Ethics*, 23(4):1213–1226.

Theurillat, J.-P., Willner, W., Fernández-González, F., Bültmann, H., Čarni, A., Gigante, D., Mucina, L., and Weber, H. (2021). International Code of Phytosociological Nomenclature. 4th edition. *Applied Vegetation Science*, 24(1):e12491. tex.copyright: © 2020 The Authors. Applied Vegetation

Science published by John Wiley & Sons Ltd on behalf of International Association for Vegetation Science.

Tomaska, L. (2007). Teaching How to Prepare a Manuscript by Means of Rewriting Published Scientific Papers. *Genetics*, 175(1):17–20. tex.copyright: Copyright © 2007 by the Genetics Society of America.

Torres, D. H. and Pruim, D. E. (2019). Scientific storytelling: A narrative strategy for scientific communicators. *Communication Teacher*, 33(2):107–111.

Toulmin, S. E. (2003). *The Uses of Argument*. Cambridge University Press.

Trivers, R. (2011). *The Folly of Fools: The Logic of Deceit and Self-Deception in Human Life*. Basic Books, New York, NY, 1st edition edition.

Turbek, S. P., Chock, T. M., Donahue, K., Havrilla, C. A., Oliverio, A. M., Polutchko, S. K., Shoemaker, L. G., and Vimercati, L. (2016). Scientific writing made easy: A step-by-step guide to undergraduate writing in the Biological Sciences. *The Bulletin of the Ecological Society of America*, 97(4):417–426.

Uhlmann, E. L., Ebersole, C. R., Chartier, C. R., Errington, T. M., Kidwell, M. C., Lai, C. K., McCarthy, R. J., Riegelman, A., Silberzahn, R., and Nosek, B. A. (2019). Scientific utopia III: Crowdsourcing science. *Perspectives on Psychological Science*, 14(5):711–733.

van Raan, A. F., Moed, H. F., and van Leeuwen, T. N. (2007). Scoping study on the use of bibliometric analysis to measure the quality of research in UK higher education institutions. Technical report.

Vellend, M. (2010). Conceptual synthesis in community ecology. *The Quarterly Review of Biology*, 85(2):183–206.

Vences, M. and Zardoya, R. (2004). Meyer: Disagreements but no misconduct. *Nature*, 431(7008):505–506.

Vonnegut, K. (1980). How to write with style. *International Paper*, 2.

Vonnegut, K. (1982). How to write with style. In *How to Use the Power of the Printed Word*. Anchor Press.

Wake, D. B. and Vredenburg, V. T. (2008). Are we in the midst of the sixth mass extinction? A view from the world of amphibians. *Proceedings of the National Academy of Sciences*, 105(Supplement 1):11466–11473.

Wang, Y., Taylor, L., Pearl, M., and Chang, L.-S. (2004). Effects of Tai Chi exercise on physical and mental health of college students. *American Journal of Chinese Medicine*, 32(3):453–459.

Williams, C.F. (2021) *Move!: The New Science of Body Over Mind*. Profile Books.

Wilke, C. O. (2019). *Fundamentals of Data Visualization: A Primer on Making Informative and Compelling Figures.* O'Reilly Media.

Woolston, C. (2016). Faking it. *Nature,* 529(7587):555–557.

Woolston, C. (2019). PhDs: The tortuous truth. *Nature,* 575(7782):403–406. tex.copyright: 2021 Nature.

Xie, Y. (2016). *Bookdown: Authoring Books and Technical Documents with R Markdown.* CRC Press.

Xie, Y., Allaire, J. J., and Grolemund, G. (2018). *R Markdown: The Definitive Guide.* CRC Press.

# Index

*Italicized* pages refer to figures and **bold** pages refer to tables.

Printed in Great Britain
by Amazon

37615487R00163